Inclusive Environments and Access to Commercial Property

This book presents and examines the challenges and compromises required to deliver inclusivity in the existing commercial-built environment and the socio-economic benefits that could result from successfully delivering it.

To illuminate the advantages of an inclusive environment to property owners, investors and service providers, the book covers the history of disability and evolution of the legislation and examines the demographics and types of disability to question the 'one size' 'blanket' approach that currently exists to providing access. Delving further into the characteristics of the commercial property sectors and individual disability-specific requirements, experienced commercial building surveyor, Adrian Tagg, analyses the contradictions in the existing legislation to establish examples of design compromise or reasonable adjustments. He seeks to contextualise public and commercial attitudes to disability and go further to demystify the term 'reasonable adjustment', which is used currently as a tool of compromise in providing access. The aim is to assess disability-specific requirements for access, as well as adopt a simplistic approach to developing access solutions to the existing built environment from a consultancy and user perspective.

Ultimately, this publication hopes to promote accessibility and inclusion from the perspective of surveyors, investors and landlords working in commercial property. It is not just targeted at those on undergraduate or post-graduate surveying courses, as well as those early career professionals undertaking their APC or post-graduate qualifications, but also at those owning or delivering goods, services and employment from commercial premises who want to make a difference.

Adrian Tagg MRICS is an Associate Professor at the University of Reading, UK. He has over 20 years of experience as a practitioner and is currently the lead for Building Surveying at the University. Adrian was awarded the title of Associate Professor in 2019 and lectures on Building Pathology, Commercial Surveying Practices as well as Inclusive Environments. His research interests concern accessibility in the built environment and was the joint winner of the 2019 UCEM Harold Samuel Research Prize, publishing *Public and Commercial Attitudes to Disability in the Built Environment* in October 2020. He retains a practice presence with a wide range of surveying experience in multi-disciplinary environments in the delivery of Technical Due Diligence to International investors making acquisitions in Europe as well as a number of UK clients and projects. Adrian is the designated author for *Due Diligence* on the RICS isurv subscription website, the sole author of the book *Technical Due Diligence and Building Surveying for Commercial Property*, Routledge.

Inclusive Environments and Access to Commercial Property

Adrian Tagg

Routledge
Taylor & Francis Group

LONDON AND NEW YORK

Designed cover image: Adrian Tagg

First published 2025
by Routledge
4 Park Square, Milton Park, Abingdon, Oxon OX14 4RN

and by Routledge
605 Third Avenue, New York, NY 10158

Routledge is an imprint of the Taylor & Francis Group, an informa business

British Library Cataloguing-in-Publication Data
A catalogue record for this book is available from the British Library

Library of Congress Cataloguing-in-Publication Data
Names: Tagg, Adrian, author.
Title: Inclusive environments and access to commercial property / Adrian Tagg.
Description: Abingdon, Oxon : Routledge, 2025. | Includes bibliographical references and index.
Identifiers: LCCN 2024018029 (print) | LCCN 2024018030 (ebook) | ISBN 9781032145532 (hardback) | ISBN 9781032137520 (paperback) | ISBN 9781003239901 (ebook)
Classification: LCC KD737 .T34 2024 (print) | LCC KD737 (ebook) | DDC 342.4108/7--dc23/eng/20240528
LC record available at https://lccn.loc.gov/2024018029
LC ebook record available at https://lccn.loc.gov/2024018030

ISBN: 978-1-032-14553-2 (hbk)
ISBN: 978-1-032-13752-0 (pbk)
ISBN: 978-1-003-23990-1 (ebk)

DOI: 10.1201/9781003239901

Typeset in Optima
by MPS Limited, Dehradun

Contents

Acknowledgements

Thank you to family, friends and colleagues for your continued support (you know who you are). Special thanks to:

Dr Geoff Cook
Olivia Tagg-Van Lany (Illustrations)

Thanks is also given to those with a disability who have responded to various research requests over the years; your voices and opinions have not gone unheard.

In memory of Sylvia Isenborghs, your love of life and determination to succeed when faced with adversity has been and always will be an inspiration.

Preface

Access to the commercial buildings that deliver goods, services and employment for those with a physical or mental impairment in the UK is mostly sub-standard. Despite legislation preventing disability discrimination being in place since 1995, there is widespread apathy as well as anger from those with a disability that their basic access rights are not afforded.

The Paralympics has proven that those with disability can do extraordinary things and compete at the highest sporting level. However, disability is not a four-year event, and there are millions of ordinary people who are prevented from accessing commercial property, goods, services and employment by the physical barriers present in the built environment. This social model of disability places an emphasis on society to remove these barriers.

This book plots the history of disability within the UK, recognising the diverse nature and definitions of both physical or mental impairments. It introduces the notion of the commercial model of disability and examines the base legislation afforded to prevent discrimination. There is a recognition that the retrospective implementation of accessibility can be challenging within the existing built environment, and in particular, to historic buildings. The text looks at both public and commercial attitudes to disability before establishing design solutions and the implementation of best practice in providing a fully inclusive built environment. While there are many nuances associated with disability and specific access needs of those wanting to engage with service providers or employers, this book is **explicitly concerned with the physical access of buildings**.

There is a sense that commercial building owners and those providing access to properties, goods, services and employment should choose opportunity over obligation to engender a fully inclusive built environment. This book is dedicated to all those with a disability or those caring with someone with a disability who struggle to access the commercial built environment. It also recognises and reaches out to those within the commercial property sector who want to make a difference; **together we can do this**.

Chapter 1

Introduction

Inclusive Environments

Inclusive environments and access to commercial property concerns the accessibility of goods and services or access to the workplace for those who recognise as having a physical or mental impairment. In plain terms, an inclusive environment seeks to provide access for all, irrespective of the wide spectrum of disability and the complex nature implementing this access within the existing built environment. It is important to differentiate that this book concerns the physical access to commercial property, goods, services and employment. It addresses the longstanding problems concerning the availability of accessible parking spaces and the paths, pavements, ramps and steps leading to the principal entrances of buildings. It analyses the concept, requirements and presence of accessibility through entrance doors, lobbies and spaces to all commercial property, including shops, restaurants, offices, hotels and transport hubs, to name a few of these applications. The book examines the challenges of vertical circulation and what it means for someone who is disabled to navigate changes in level, as well as the location and specification of a range of facilities, including toilets. What this book does not assess is the role of service providers and what they need to do to make their service accessible. An example of this accessibility is with a theatre or cinema. This text will identify the physical components that make up the entrance, doors, facilities or circulation space, as well as recommending methods to facilitate access. It will not analyse if the performance or film is in an accessible format, say with the use of subtitles for the deaf or audio description for the blind.

There is no doubt a need to recognise that physical access and the accessible products delivered by service providers both need to be considered for a holistic approach to full inclusivity. The provision of an inclusive environment is the starting point, and once it has been established that there are no or few reasons why buildings cannot be accessible, service providers cannot hide behind the notion that they have to make significant alteration to their properties to facilitate this accessibility. It is too easy to adopt the path of least resistance when considering access to commercial property, goods, services or employment. To be frank, and despite over 20 years of legislation preventing disability discrimination, UK commercial properties remain largely inaccessible. Claims of inclusivity are too often paying lip service or appear a box-ticking exercise to fulfil an accreditation or comply in part with a notion of corporate social responsibility on this issue. This is evident with the perceptions of disability and what it is to be disabled.

DOI: 10.1201/9781003239901-1

Historically, there has been the misconception that disabled access concerns primarily wheelchair users and access revolving around the provision of a step-free environment. However, wheelchair users are part of a much wider demographic who identify as having a disability and who have a differing access need. Added to this issue is the fact that those who recognise as being disabled often can have more than one physical or mental impairment.

It is widely acknowledged that disability is a global issue that is prevalent in all walks of life irrespective of wealth, culture, custom and race. Historically in the UK, there is evidence of persecution and discrimination to those with a physical or mental impairment; this topic is explored fully in Chapter 2. Despite this discrimination, there have always been aspects of society willing to reach out and offer help.

Notwithstanding this problem, it was not until the latter part of the 20th century in the UK that disability was formally recognised in a legal context, and it has been more than two decades since the much-heralded *Disability Discrimination Act 1995*. This legislation placed an obligation on commercial property owners, as well as those delivering goods and services, to make these reasonably accessible to those with disability.

Therefore, it is hard to conceive that in 2017, there should be a report titled *Being disabled in Britain: A journey less equal*[1] considering this was 5 years after the 2012 Paralympics in London. This event, which was widely recognised as being the most successful Paralympics of all time, setting legacy goals specifically to reduce discrimination and improve opportunity for those with a disability.[2]

Despite the presence of a legal framework, enhanced public awareness and high-profile sporting events showcasing disability, it is perplexing and a reality check that those with a physical or mental impairment still feel discrimination. This reality is evident in research periodically undertaken by the charity Scope, which shows a widening of the 'perception gap'.[3] In simple terms, those with a disability perceive they are discriminated against more than the perception of disability discrimination as viewed by those without disability. One reason for this gap may be attributable to 'experiential understanding' as identified in a report published by the University College of Estate Management.[4] This research undertaken in to commercial attitudes to disability in the built environment validates the presence of a significant perception gap. This concerns specifically the accessibility between those operating or advising on commercial property delivering goods and services compared to those with disability accessing goods and services themselves.

While the 'perception' of disability discrimination may be considered subjective and really only felt by those struggling to gain access to goods and services, legal obligation should be far more objective. Guidance exists to direct commercial organisations, service providers and employers to deliver accessibility on the basis of making the necessary 'reasonable adjustment' to facilitate it. However, this can seem quite confusing, and the very nature of the term 'reasonable' makes this open to interpretation. Furthermore, as seen with legal obligations, these are often treated with a degree of distain with many service providers seeking to conform to the legal minimum. This 'race to the bottom' neglects the commercial benefit that an inclusive environment can bring. Unlike most other legal requirements associated with the built environment, the implementation of above legal minimum standards has the potential the enhance commercial viability. Put into context, in building laws, there is little commercial advantage in installing a one-hour fire door to property if the legal requirement is for 30-minute fire resistance.

However, providing above minimum legal compliance regarding disabled users enhances accessibility, which has the potential to increase use or sales of goods and services. Abandoning the 'fire door' approach to inclusivity in the commercial built environment is a win-win for business and society.

Background

Disability has no 'boundaries' as it does not differentiate between rich and poor or those within the different sectors of society. It crosses the 'frontiers' of age, race, gender, culture and religion. Globally disability is defined and recognised in accordance with the United Nations and World Health Organisation, although the treatment and support of those with disabilities varies significantly between countries, cultures and even individuals. The UK may be considered to have a progressive approach to disability, and this is recognised within *The Equality Act 2010*. According to the Act; disability is defined as *physical or mental impairment that has a 'substantial' and 'long-term' negative effect on a person's ability to do normal daily activities.*[5] The Equality Act also places an obligation on service providers and employers to make *reasonable adjustment* to ensure they do not discriminate against those with disability. It therefore appears an obvious requirement for those supplying goods and services or employing those with disability, from premises within the built environment, to make the necessary adaptive changes to their properties. However, the legal framework is relatively new and as such, is the subject of ongoing legal cases which should begin to shape the application of the law. The term *reasonable adjustment* has a degree of ambiguity for auditors, designers and users, this too is the subject of legal interpretation which will be developed as challenges are made under the legislation.

There are a large amount of legal prescriptions associated with the design, planning, construction and occupation of the built environment, but relatively little is devoted to the provision of access for those with disability. For new-build properties or those the subject of significant renovation where it is necessary to seek building regulations approval, *Approved Document Part M* (2010) provides the relevant detail of what is required to deliver access. However, a significant quantity of commercial premises delivering goods and services in the UK comprise existing buildings. Considering that commercialisation as we might know it today began in the highstreets of the UK in the Victorian era, there are close to 160 years of commercial property that was never initially designed to comply with accessibility legislation. Clearly, much of this has gone through historical renovation or change of use, including adaptation for those with disability depending upon the legislation or guidance at the time. Despite this, there is an appearance of key building elements or components that form part of the structure or envelope which simply cannot be altered without major structure intervention. This is probably why the term 'reasonable adjustment' was conceived by lawmakers concerned with forcing the near-impossible retrospective alteration of existing properties to comply with current legislation. Accordingly, there will always be a need for compromise, but why bother if all we are talking about is a number of wheelchair users that can be counted in hundreds of thousands compared to a population measured in tens of millions? The truth is that as much as there is a complexity and diversity concerning the existing built environment, this is matched by a large percentage of the UK population with very real and different access needs. In this sense, the term 'reasonable adjustment' serves only to exacerbate the nature of compromise.

Approximately one in four people in the UK are disabled,[6] and it is not surprising that within the category of those considered pensioners (the over 65s), over 40% identify as having a physical or mental impairment. Disability has the potential to affect everyone either directly or indirectly, even more so with age. Prior to the *Equality Act 2010*, the provision of access to the built environment was governed by *The Disability Discrimination Act 1995*, which was revised in 2005 and eventually replaced by *The Equality Act 2010*. Effectively there has been a significant quantity of years devoted to 'progressive' legislation obliging service providers and employers to facilitate access and it is not unreasonable to question why those with disability should still struggle to access goods or services on a daily basis? There is a need to establish if goods and service providers or employers take accessibility seriously. Furthermore, is the implementation of 'reasonable adjustment' or simply if the compromises implemented are insufficient and the term 'reasonable' too open to interpretation.

Into the 21st century, public perception or attitudes to disability appears to be changing with the much-heralded success of the Paralympics held in London in 2012. The *We're the Superhumans*[7] trailer for Channel 4's coverage of the 2016 Rio Paralympics was watched 23 million times on *Facebook* alone within the first five days of going live and is evidence that there as huge public interest in disability sports. There appear continued positive attitudes to the integration of disability with most governing sports bodies in the UK, Europe, and across the globe. Despite this prevalence, it is not wholly clear what public perception exists concerning perhaps the more mundane, everyday issues affecting those with disability.

Contradicting the appearance of positive public attitude to disability, as encompassed with the Paralympics, are the 'struggles' of those living with disability on a daily basis. Disability is not like the Paralympics, a once every four years occurrence. The report titled *Being disabled in Britain: A journey less equal*[8] suggests there remain significant short comings with the provision of fully inclusive environments. Contained within the report is a statement from the chair describing those living with a disability are treated as *second-class citizens*. This indicates that despite all the positive publicity and 'warmth' towards disabled athletes, this does not appear to have filtered down into the wider disabled community. There appears a disconnect between the 'one nation' approach to inclusivity in sport and the more general needs of the disabled for the accessibility of goods and services or employment within the built environment.

To develop an understanding on how to achieve an inclusive environment, it is necessary to frame the discussion around the term 'disability' with a historical analysis of disability in the UK. This is important because, for many observers, disability focus is often in the forefront of society with high-profile events such as the previously mentioned Paralympics. Accordingly, the spotlight on disability has perhaps only been evident since the late 1990s, while in reality, there are key records of disability stretching back many centuries. It has no doubt been a struggle for those with a physical or mental impairment, which is far from over; however, through historical analysis, it is evident that there have been those who have reached out to help. Accordingly, wounded soldiers returning from the 'Great War' (1914–1918) prompted focus on disability and may be seen as the beginning of the current 'journey' to where we are today. However, progress has not been 'linear', and with an increase in the percentage of the population recognising as having a physical or mental impairment, the journey appears far from complete. There is open criticism that there is much more that could be done to promote access equality, as well as too many organisations paying 'lip service' to *diversity and inclusion* (D&I) or *Corporate Social Responsibility* (CSR). To plot the journey and attempt to influence the direction of travel regarding inclusivity in the built environment, it is necessary to establish the factual past, present and future. Determining the events that have shaped society's awareness of disability and the key events that have influenced decisions on implementing change are critical in driving the agenda on improving accessibility for the future.

Notes

1 Being disabled in Britain: A journey less equal | Equality and Human Rights Commission (equalityhumanrights.com)
2 Inspired by 2012: The legacy from the Olympic and Paralympic Games (publishing.service. gov.uk)
3 Disability Perception Gap | Disability charity Scope UK
4 Public-and-Commercial-Attitudes-to-Disability-in-the-BE.pdf (ucem.ac.uk)
5 Equality Act 2010 (legislation.gov.uk)
6 CBP-9602.pdf (parliament.uk)
7 We're The Superhumans | Rio Paralympics 2016 Trailer – YouTube
8 Being disabled in Britain: A journey less equal | Equality and Human Rights Commission (equalityhumanrights.com)

References

Channel 4 Entertainment. (2016). *We're the superhumans/Rio Paralympics 2016 trailer.* [online] Available at: https://www.youtube.com/watch?v=IocLkk3aYlk [Accessed 14 January 2024].
Dixon, Smith and Touchet. (2018). *Disability perception gap – Policy report.* Scope. [online] Available at: https://www.scope.org.uk/campaigns/disability-perception-gap/ [Accessed 14 January 2024].
Equality and Human Rights Commission. (2017). *Being disabled in Britain: A journey less equal.* [online] Available at: https://www.equalityhumanrights.com/sites/default/files/being-disabled-in-britain.pdf [Accessed 14 January 2024].
Gov.UK. (2015). *Inspired by 2012: The legacy from the Olympic and Paralympic Games – Forth Annual Report Summer 2016.* [online] Available at: https://assets.publishing.service.gov.uk/government/uploads/system/uploads/attachment_data/file/544197/1776-E_Legacy_Report_2016_ACCESSIBLE.pdf [Accessed 14 January 2024].
Gov.UK (n.d). *The Equality Act 2010.* [online] Available at: https://www.legislation.gov.uk/ukpga/2010/15/contents [Accessed 14 January 2024].
Kirk-Wade. (2023). *UK disability statistics: Prevalence and life experiences.* Research Briefing – House of Commons Library. [online] Available at: https://researchbriefings.files.parliament.uk/documents/CBP-9602/CBP-9602.pdf [Accessed 14 January 2024].
Tagg. (2020). *Public and commercial attitudes to disability in the built environment.* [online] Available at: https://www.ucem.ac.uk/wp-content/uploads/2020/10/Public-and-Commercial-Attitudes-to-Disability-in-the-BE.pdf [Accessed 14 January 2024].

Chapter 2

History of Disability in the UK

Introduction

For many, the notion of disability and, in particular, the attitude of society, is derived from what we know, what we experience or what we are taught. At the forefront is the adoption of attitudes to disability in the sporting environment, as evident in the Paralympics or other national and global events. Accordingly, it is hard to imagine that pre-1995, there appeared so little emphasis on such inclusion. However, the awareness of those with a physical or mental impairment certainly existed long before the 20th century and the advent of *The Disability Discrimination Act 1995*. Regrettably, and aligned with many other minority or marginalised sectors of society, there appears to have been a misunderstanding as well as mistrust of those with disabilities. It is impossible to comprehend the treatment received by those with disabilities or even the terminology used to refer to those with physical or mental impairment.

Much the same as with race, colour, nationality, ethnic or national origin along with religion or belief, sex, sexual orientation and other 'protected characteristics' defined by *The Equality Act 2010*, there is an acceptance of the need to recognise disability. While disability awareness appears on face value to have been widely adopted by society and is also embraced by a wide variety of organisations, it is confusing to acknowledge that there is a 'perception gap' between those with disability and those without. While it should be accepted that there has been and continues to be bigotry against those with 'protected characteristics', there is legal protection afforded. Acceptance of those covered by the protected characteristics in *The Equality Act 2010* comes from within the individuals who make up 'society' as a whole. However, disability stands alone in the notion that discrimination cannot be solely resolved with a 'hearts and minds' approach. The reality is that there will always be a need to make physical changes to the built environment to in part solve discrimination, and while influencing hearts and minds is relatively low cost, implementing alterations to buildings costs money and is again likely to have contributed to the creation of the term 'reasonably adjustment'.

Pre-20th century, there appears very little acceptance of the need or willingness to accommodate disability in society. Despite this fact, there has always been an aspect of society willing to reach out and help those with a physical or mental impairment.

DOI: 10.1201/9781003239901-2

Medieval England (1050–1485)

There is a somewhat brutal historical reference to the sick and those with impairments in Medieval England, as examined by Historic England[1] with many terms including 'leper', 'lame', 'lunatick', 'natural fool' and 'creple' used to describe those with impairments. Although perhaps 'normalised' in medieval times, such attitudes were also noted to be present in the early 20th century, albeit reserved for extremist views of a relatively small part of society. However, in Medieval England, the power and influence of the Church was such that disability was seen in contradicting terms. For some, it was seen as being a form of punishment for sin; for others, a sense of purgatory or suffering on earth making them closer to death and ultimately heaven. A sense of shame on those with disability meant that they were shunned and often looked after by friends or family, although those devoid of support resorted to begging. Ironically, it was the Church, through monks and nuns, who cared for the disabled and sick with a sense of Christian duty. During this period, Historic England noted that a network of hospitals and Almshouses appeared for the care of the sick, disabled and elderly (Fig 2.1).

Echoing the notion that disability has little 'respect' for wealth, culture or religion, the King of England between 1483 and 1485 was Richard III, who had a deformed spine as a result of 'scoliosis',[2] Shakespeare's observation of the king in his play Richard III includes the description of the king as *Deformed, unfinish'd,* and the play portrays his frail reign. This is one of the earliest references to disability appearing in a text 'published' within the public domain. It is poignant that this reference also illustrates disability from a negative perspective.

Figure 2.1 (a) and (b) Almshouse in Malmesbury (Wiltshire) including inscription detailing benevolent donation to the poor of the Almshouse.

There is an irony that one of the most coveted roles written by Shakespeare has historically been played by able-bodied actors who use the term 'cripping up' when preparing for the role and that of other disabled characters.[3] Despite this, and in 2022, approximately 430 years after writing the play, the Royal Shakespeare company cast disabled actor Arthur Hughes as the first person with a disability to play the lead role. Despite what can only be described as a positive example of inclusivity and irrespective of historical evidence concerning the disability of King Richard III, there is a perception of weakness associated with the king that might go further than the physical disability. There is much debate concerning the historical portrayal of the disabled monarch; esteemed expert Prof Michael Dobson, director of the Shakespeare Institute, examines the various 'myths' in his article *Richard III: The real king of history, or marvellous theatrical villain?*[4] Dobson concludes that although not totally 'innocent', the portrayal of Richard by Shakespeare is largely created as a dramatic theatrical representation by a celebrated author. This, in keeping with Shakespeare's signature body of works, has ultimately resulted in a misrepresentation of the actual king of England. It could be additionally argued that this in itself creates a gap in the perception of disability to the reader, far removed from the reality. Although perhaps identifying as one of the earliest representations of disability in the published public domain, there is a poignancy that this simultaneously appears a 16th century misrepresentation of disability. Misrepresentation of disability and particularly a *perception gap* is still a significant barrier facing those with physical or mental impairments in the 21st century.

Henry VIII to Victorian Britain

The influence of the Catholic church was diminished following Richard's successor Henry VIII and his dissolution of the monasteries and the support network for the provision of care for the disabled.[5] The dissolution of the monasteries was a policy that would have the most direct impact of so many of his subjects,[6] although there is no information or data detailing the prevalence of disability within communities of the Middle Ages. Despite this lack, it is documented in research undertaken by Historic England[7] that there was a petition in 1538 calling for the re-opening of closed hospitals, and to an extent, this eventually succeeded. It is not clear if the strength of public opinion influenced the decision or, as alluded to, was the sense of offence caused to passers-by having to be exposed or witness those with disability in the street. In their research, Historic England attributed the following powerful quote in the petitioning to reopen hospitals:

> the miserable people lyeing in the streete, offending every clene person passing by the way

This shocking statement again illustrates the perception of disability in the community from a very negative perspective and does little to suggest genuine concern or care. Such polarised perspectives on disability appear throughout the passage of time and further emphasises the perception gap between what it is to be disabled and how disability is viewed from those without a physical or mental impairment.

Despite the petition to reopen the hospitals, there was little evidence to suggest any tangible changes in the following 30 years. Ultimately, those with disability were likely to be amongst the most affected by the Reformation due to their reliance on the church for help. In general terms, society has come a long way in many respects since the Middle Ages, but concerning disability, there is still the notion that those considered less fortunate are most affected by negative change. The withdrawal of support for those with disability has been replicated throughout the centuries. This is even apparent with policies of austerity adopted by the UK government during the early 21st century, despite the much-heralded success of the London Paralympics in 2012.

History has taught us that while there have been perpetrators of discrimination, there have also been those willing to support and care for those with physical and mental impairment. Such paradox is also evident post Henry VIII with text contained in a handbook for justices of the peace. This advised:

The person naturally disabled … not being able to work … are to be provided for by the overseers [of the poor] of necessary relief and are to have allowances proportional[8]

This perhaps was one of the more formal, 'published' attempts to positively address disability with a structured 'policy' as opposed to the benevolent, 'Christian' approach adopted by the church. Advances on the sentiments alluded to above are evident during the 17th, 18th and the early 19th centuries.

There is evidence of changing attitudes, and accordingly, those with disability were observed to have experienced a 'misfortune' as opposed to a punishment from God. As such, those with disability received a level of charity and were cared for in hospitals. This newfound approach to recognising and providing care for those with disability suggested that public opinion was beginning to change again towards the end of the 18th century. However, in reality, opinions concerning the recognition of disability and providing support have always been polarised; it often appears either *all or nothing,* and this has shown to be the case from Medieval Britain through to the 21st century. The industrial revolution saw the construction of civic buildings, including hospitals. This resulted in public opinion that the institutions were the 'right place' for disabled people, although it is acknowledged that probably fewer than 10,000 people amongst a population of nine million resided in such institutions.[9]

It is acknowledged that the 19th century saw a boom in buildings designed for people with disability, and, as reported in *The Builder Magazine* 1892, groups of buildings befitting a town located in the adjacent secluded countryside, which were asylums.[10] Research by Historic England illustrates an apparent eagerness in the Victorian era for the disabled to populate workhouses. Despite this, it can be critically appraised that this was not entirely for the right reasons. According to Historic England, it was also an attempt to prevent those *shirkers and scroungers* disguising themselves as handicapped and receiving financial assistance and aid whilst remaining at home. As detailed by Historic England, the Poor Law Amendment Act of 1834 encouraged the industrious effort to construct and fill asylums and workhouses up and down the country. The attitude to recognise yet banish those with a physical or mental impairment might suggest that the Victorians were conforming to the all-or-nothing approach to disability of their predecessors. Clearly, there is evidence that the Victorians were good at recognising the challenges of disability since in a generation of inventors and engineers, disability might have been perceived as a 'problem'. However, unlike the imposition of great feats of engineering associated with profound need to solve infrastructure or engineering problems, disability needs to be confronted with integration, not segregation in society.

There is widespread historical opinion that workhouses were indeed dreadful places that incarcerated the poor and impoverished, often punishing them for reasons beyond their control. There was a feeling amongst wealthy and middle-class Victorians that they were 'paying' for the poor, sick and disabled, and accordingly, the building of workhouses was a very effective way to remove them from the public domain. Such attempts to address this issue, and in particular, the context of disability appears a repeat of the justification to petition the re-opening of hospitals some 300 years earlier.

The need to recognise disability was brought into sharp focus in the early 20th century with the onset of the first of two World Wars. The consequence of conflict would force society to deal with disability arising from the very manmade brutality of war. The development of relatively sophisticated munitions saw tanks replace cavalry, as well as bullets and shells, displacing the sword and dagger. The consequences of weapons designed to kill at best or indiscriminately maim in the worst case had a profound effect on the recognition of disability.

Post World War 1 (WW1) to 1995

Post WW1, an estimated eight million soldiers returned home permanently disabled after the conflict and of these, two million were British servicemen.[11] Victory in the *Great War* resulted in a generation of formerly physically fit men returning from the battlefields with long-term physical and mental impairments affecting normal daily activities. The horrific injuries caused by conflict included visible disabilities such as the full or partial loss of limbs and blindness as well as breathing difficulties caused by the crudity of early chemical warfare. There were those deemed to be suffering from 'shell shock', which is now recognised today as *Post-traumatic Stress Disorder* (PTSD).

Faced with the reality of honouring those victorious in battle with the notion to construct *homes fit for heroes*, as detailed in a speech by Lloyd George the day after Armistice Day, *attitudes had to change.*[12] Accordingly, changes were made to the building and development of housing (Hasted, 2016). The plight of those living with a disability in post-war Britain was recognised by the lead architect and town planner T H Mawson. He was instrumental in the development of disabled 'friendly' properties, stating that life was a *struggle on the part of the crippled man with those who are able-bodied* (Mawson, 1918). Employers were encouraged to take on disabled ex-servicemen, who were housed in specific properties from single cottages to entire bespoke villages. Parallel to this support, civilians with disability were encouraged to live in rural colonies. This is perhaps an attempt at integration, but the published findings of the census of 1921 and accessible 100 years later highlights the frustrations at the lack of action on disability in comparison to the political rhetoric.[13] One respondent wrote words to the effect to stop talking and start building in respect of the pledge of the then liberal prime minister to build one million properties for those returning from war. Others wrote that they were *ruined by war,* and perhaps one of the more profound responses was from an army officer about why he had broken the census rules and typed his answers instead of handwriting them. He detailed in the response that this was because he had lost part of his right hand in the conflict. The use of a typewriter to format these specific answers in the context of completing the 1921 census may have been frowned upon in 1921, but in the 21st century, this would be seen a completely acceptable as a 'reasonable adjustment'.

The end of the *Great War* is seen by some as a watershed moment and, as discussed in a blog by Historic England,[14] there was some pioneering technology concerning the development of prosthetic limbs; the use of aluminium components with adjustable joints replaced the cumbersome timber prosthetics. There were also advances in plastic surgery, which is remarkable since this was more than 100 years ago, with some of the work on facial reconstruction incredible. There was a willingness within aspects of society to provide places of work for those with disability, and organisations like *The Poppy Factory*, founded by Major George Howsen MC, were established to provide work for disabled veterans. Founded in 1922, *Thc Poppy Factory* celebrated its Centenary in 2022; along with the *Royal British Legion,* it still provides support for disabled servicemen. Homes were built and in the terms of construction technology, the interwar period is seen by some as the era that saw the 'birth of the bungalow'. Although seen as less fashionable and shunned by today's property developers, who are engendered to build vertically to maximise plot-to-profit ratio, the bungalow eliminated one of the biggest barriers to accessibility, and that is vertical circulation.

Despite the suggestion of positivity towards those with a physical or mental impairment in post-WW1 Britain, it was observed that those with severe disabilities were discriminated against in the workplace. Furthermore, a lack of understanding on the mental trauma referred to today as *PTSD* resulted in some barbaric treatments, such as electrocution or solitary confinement to deliver a 'cure'.[15] There is no doubt a wide spectrum of what constitutes disability, and this topic will be discussed fully in this book; however, discrimination is discrimination, irrespective of where it lies on a spectrum. Regrettably, the notion of *Love thy neighbour … .unless they are ….* is an attitude that often prevails with disability. While it is not a direct science, there is a school of thought that for every positive, there is a negative in respect of those reaching out to support those with a disability. It is therefore necessary to point out that Post World War One Britain relied mostly on charity or voluntary commitments in disability support. Not obvious in the early 20th century was evidence or research alluding to the perception of disability discrimination amongst those with a physical or mental impairment. However, there is published data for the 21st century that confirms those with a disability perceive the presence of discrimination significantly more than those who don't have a disability. In plain terms, only those with a disability or caring for someone with a disability have experiential understanding of discrimination. In many ways, it is likely that those with disability were significantly more discriminated against in the interwar years despite the perception of 'change' engendered by those returning from conflict.

The Great War had taken its toll on the population of the UK with the loss of life as well as a significant drop off in the birth rate the population dipped significantly just as the UK and rest of the world entered serious economic downturn. Despite the perceived advances in attitudes to disability post-WW1, this was undermined by the growth in populism of the eugenics movement in the late 19th and early 20th centuries. In 1930, Julian Huxley, Chairman of the Eugenics Society wrote *every defective man, woman and child is a burden.*[16] It is not known if such a forthright view reflected what was widely felt in society at that time. The notion that natural selection or survival of the fittest is valid meant there was little empathy evident by supporters of such theories towards those with disability. It is a reminder of the often-polarised opinions concerning disability, however, this time, with an emphasis placed more on the social economic issues levied as an argument against providing support. Indeed, nearly 80 years later, as the world faced economic uncertainty and a global banking crisis in 2008, similar opinions re surfaced. With the background of economic gloom and subsequent recession in the UK, a speech by the then chancellor of Exchequer George Osborne suggested benefit culture in the country was a lifestyle choice. While the speech did not specifically refer to those supported by disability benefits, it was observed at the time that using an umbrella statement to criticise those on benefits also inferred to those with a physical or mental impairment could be potentially damaging. Disability is far from a lifestyle choice, and this was put into sharper focus as war raged again across the globe in the mid-20th century.

It is estimated that WW1 killed approximately 880,000 service personnel and resulted in over 40,000 amputees,[17] the WW2 caused even more death and destruction. Despite this, the 1940s and 1950s saw significant advances in the recognition and treatment of those with disability. Post WW2, there appeared a revolution in progressive legislation with the introduction of the 1944 Disability Employment Act, which was directed at disabled participation in the workplace and rehabilitation.[18] This was also an important period in history concerning the welfare state and the 'birth' of the National Health Service (NHS) in 1948. One of the founding principles of the NHS was that free health care for everyone, and the NHS is still in existence today. Currently, the NHS essentially offers health care free at source, with the current governing principles including to *make sure nobody is excluded, discriminated against or left behind* being that *It is available to all irrespective of gender, race, disability, age, sexual orientation, religion, belief, gender reassignment, pregnancy and maternity or marital or civil partnership status.*[19] This very statement embodies the spirit of true inclusivity and aligns in principles contained within the Equality Act 2010.

Another important poignant event occurring in 1948 was the Stoke Manderville Games, which were widely recognised as the forerunner to the current Paralympic Games. Initially hosted at the Stoke Manderville Hospital, staff and inpatients held a series of sporting challenges, including archery.[20] Also in 1948 was the passing of National Assistance Act, which had a section dedicated to the welfare of those with disability.

According to Historic England,[21] disabled people did not remain passive in the 1940s and 50s with the formation of the charity SCOPE in 1952, a driving force for change. Indeed, the second half of the 20th century would see significant changes in the attitudes to disability in the UK, as well as globally. Inspired by the civil rights movement in the USA in the 1960s and 70s, disabled groups in the UK were empowered to act against inequality, discrimination and poor access. This was triggered from having seen the success of activists in the USA in achieving social and political change. Accordingly, a *social rather than medical model of disability emerged,*[22] and this emanated from a 'ground breaking' publication *Fundamentals of Disability* published by *The Union* in 1976. According to the late Professor Mike Oliver, known as *the father of the social model of disability,* it is society and not an individuals' impairment that creates disability.[23] This very important shift in opinion towards the responsibility of society in shaping disability will be opened up for further discussion in Chapter 3. The application of reasonable adjustment to address societal disability is covered in greater depth throughout this book.

Described as the *Magna Carta for this disabled*,[24] the Chronically Sick and Disabled Persons Act 1970 (CSDPA70) was seen as being an influential piece of legislation in the drive for equality. Occurring at a time of significant 'change' in attitudes towards disability, the CSDPA70 included for the following provisions:

- Education and support at home.
- Access to public buildings.
- Disabled badges.
- Representation on public bodies.
- Segregation in hospitals.
- War pensions.

Evolving out of CSDPA70 was the Disabled Persons Act 1986 (DPA86), which placed further emphasis on the assessment of disabled people in regard of their needs for care. The surge or swell in momentum for disability rights saw protests on the streets in the UK akin to the 1960s civil rights demonstrations in the US, with many disabled protesters risking arrest in the early 1990s.[25] They clamoured for their right to equality and to an end of perceived discrimination.

Revolution to Evolution

The first bespoke legislation directly dealing with discrimination associated with accessibility was the Disability Discrimination Act 1995. This set a clear definition of disability, and more importantly, obliged commercial building owners or service providers to ensure their goods and services were accessible. Building Regulations, *Approved Document Part M,* has since been seen as the default position regarding minimal legal compliance; this document, as well as other key design guidance notes, will be analysed fully in Chapter 4. It should be acknowledged that building regulations are primarily concerned with approved standards for new build and renovation. The adoption of this as benchmark to ascertain compliance for access in the context existing buildings is complex, particularly when dealing with historic or listed properties. This is also opened up for discussion in several of the following chapters. Concerning the implementation of the DDA1995, discrimination against disabled people has been supposedly been legally enforceable for over 20 years. Since October 2004, those in charge of buildings with public access have had to undertake reasonable measures to ensure the provision of access to those with disability. However, there exists much criticism concerning the act and the real-life experiences of those accessing commercial properties, goods, services and employment.

It has been concluded that despite legal provision being in existence for 20 years, many disabled people are still excluded from full participation in society.[26] This was further echoed by The Women and Equalities Committee, who stated in their 2017 report that disabled people still face challenges as basic as trying to access *public and commercial buildings without step free access.*[27] Accordingly, questions need be asked about why this can still be the case after more than 20 years of legal application and an apparent positive change in public awareness to disability.

Showcasing Disability

As discussed, when plotting the time line on key historical events concerning disability, the Paralympic Games are fundamental in raising the profile of disability. The current Paralympic movement was initially conceived in 1948 to involve wounded veterans in the London Olympics. Originally named 'the Stoke Mandeville Games', the Games changed to a format that we now know as the 'Paralympic Games' in 1960.[28] The success of the games in raising the profile of disability has been well documented. The 2012 London Paralympic Games were seen as a landmark event for disability sport; these games have also been hailed as the greatest Paralympic success of all time, *broken all records*[29] with 2.7 million tickets sold. The commercialisation of the Games was seen as vital for its financially viability; this resulted in brands such as Sainsbury's and BT turning Paralympians into *household names,* as stated by Paul Deighton.[30]

The Games captured the imagination of the wider public, and as a broadcasting spectacle, Channel 4 paid a reported £9 million to have the rights to broadcast the Games, deemed a major commercial success. However, the real success of the London Paralympics as a legacy for the perception and treatment of those with disability is debatable. The charity for the disabled, Scope, stated in a blog in 2013 that *81% of disabled people say that attitudes towards them haven't improved in the last twelve months.* This separation between public attitudes to Paralympians and non-Paralympians was confirmed by Sophie Christiansen. As a Paralympic three-time gold medal winner, she highlighted the *huge gap* between societal perceptions of Paralympians and the *rest of the disabled community.*[31] Furthermore, an article written by the blind and former home secretary David Blunkett, published in the Independent, highlighted the contradictions in public perception. Titled *The public's perception of disabled people needs to change – we're not just Paralympians or Scroungers,* the article questions reform in the welfare state and highlights the feelings of admiration for Paralympians, contrasted with a mixture of sympathy or resentment for those receiving financial support. Telling is the discussion that public opinion sides with those 'deserving' or 'non deserving' while the voices of millions of disabled people want support to participate in their community to contribute to their own well-being.[32]

While the article emphasised that those with disability primarily want help to help themselves, research by the charity Scope on the perceptions of disability identified that 75% of the sample population perceive that 75% of those with disability require care some or most of the time.[33] This idea was expressed in the *Disability perception gap – Policy report* published by Scope in 2018, which also highlights the view that 32% of those with a disability feel there's a lot of prejudice against them. This is only 5% down from the 37% expressing this belief in 2000. The 'gap' is evident in the views of non-disabled people who feel prejudice against disabled people has dropped to 22%. This is significant in highlighting the difference of perception from those experiencing disability to those observing it and is one of the many points of 'disconnection'. As alluded to on several occasions, attitudes towards disability are often polarised, and this was also post-2012 Paralympics concerning the social impact of the Games. From a built-environment perspective, Trudi Elliot of the Royal Town Planning Institute described the 2012 Games as *the shining example of inclusive planning and delivery,* However, Sophie Christiansen's criticism of the Games, their influence in altering public perceptions of disability, was commended. Despite detailing 'the gap' between Paralympians and the disabled community, Ms Christiansen described how the Games *not only inspired a generation, it challenged the ideas of a generation.* Ms Christiansen's comments should be contextualised with the opinion of the government, who reported that 70% of the *British public feel attitudes towards disabled people have improved since the London Paralympic Games in 2012.*[34]

The increased public awareness and apparent warmth to disability sport has further developed into the *Invictus Games,* which is for disabled service personnel with the aim of showing that there is a *life beyond disability,*[35] The notion that disability is not limiting is reflected in the Games' slogan, *We Came. We Saw. We're Unconquered.* Such is the importance of these sporting events that it is stated that *Paralympic sport acts as an agent for change to break down social barriers of discrimination for persons with a disability.*[36] This contemporary opinion is essential in supporting the idea that the Paralympic Games successfully breaks down the barriers placed in front of disabled people, as described in the *Social Model of Disability.*[37]

The high-profile coverage of the Paralympics appears to have engendered a sense of national pride or admiration for Paralympians. This is embodied by their *superhuman* qualities along with talent and drive to overcome adversity and be the best. The existing published literature on this subject suggests that public opinion and admiration is reserved for the stadia or sports arenas. Little spills over into the recognition of people with extraordinary adversity attempting to live very ordinary lives. There is a sense of isolation in which the general public views the achievements of Paralympians and the need of those living with disability. Historically there has been a progressive recognition and definition of disability that has been showcased on the highest possible stage. Lagging slightly behind the universal recognition of disability are the legal frameworks designed prevent discrimination and promote inclusivity within the built environment.

Importantly, during a period where disabled athletes have had the highest possible profile with the 2012 and 2016 Paralympics, this has coincided with austerity in the UK. Triggered by the banking crisis of the late 2000s and subsequent recession, the UK government undertook a radical reform of the social welfare sector. As a consequence of government cost saving, the Disability Living Allowance was replaced by the Personal Independence Payment. This has evoked controversy with claimants requiring constant review and assessment. Some have argued that this is a backwards step in government policy, whilst others have argued it is necessary to bring the [then] current £12 billion spent on benefits under control. According to a report published by Disability Rights UK, the disabled are hardest hit by reform. In their online article dated the 15th July 2019, they claim that disabled people are four times worse off financially than non-disabled people as a result of welfare reform.[38] The real implication of welfare cuts appears evident in data published by the Joseph Rowntree Foundation in 2020,[39] which show an initial decrease in disabled poverty rates from 37% in 1999/00 to 28% in 2011/12. This has currently risen to 30% in 2017/18, which is reversing the trend. In contrast, the poverty rates for families without disability is 19 percentage points lower, which is evidence of a significant 'gap' directly related to disability. Helen Barnard, writing for the Joseph Rowntree Foundation in 2019, detailed that there is gap established to represent 3/10 disabled persons living in poverty compared to 2/10 non-disabled persons. The presence of the term 'gap' is not uncommon in the context of comparisons between disabled and non-disabled persons; this idea is explored further in the *Disability perception gap – report* by Scope.

Concerning the historical context of disability in the built environment, it has been established that disability has been documented for many centuries. There is evidence that those with disability have received care and support in varying degrees, although currently there appears polarising arguments that not enough is being done to recognise and support disability in the community in the 21st century. To understand the significance of this research, it is necessary to understand clearly establish the types and nuances of different disabilities, as well as the prevalence of disability within the UK population.

Notes

1 historicengland.org.uk
2 Scoliosis | Richard III: Discovery and identification | University of Leicester
3 Arthur Hughes: First disabled Richard III is 'big gesture' from RSC - BBC News
4 Richard III: The real king of history, or marvellous theatrical villain? - University of Birmingham
5 Disability from 1485–1660 | Historic England
6 Rex, R. (2006). Henry VIII and the English Reformation. 2nd ed. Basingstoke: Palgrave Macmillan, p. 45.
7 Disability from 1485–1660 | Historic England
8 Branson, J. and Miller, D. (2002). Damned for their difference: The cultural construction of deaf people as disabled. Gallaudet University Press, p. 7
9 The lives of people with disabilities in 18th century England | Historic England
10 Disability in the 19th century | Historic England
11 Cohen, D. (2001). The war come home. Berkeley: University of California Press.
12 Disability in the early 20th century 1914–1945 | Historic England
13 Census 1921–100-year-old secrets revealed – BBC News
14 Home from the war: What happened to disabled First World War veterans – The Historic England Blog (heritagecalling.com)
15 Home from the war: What happened to disabled First World War veterans – The Historic England Blog (heritagecalling.com)
16 Disability in the early 20th century 1914–1945 | Historic England
17 The impact of the First World War | Historic England
18 Barnes-policy-and-practice-1.pdf (leeds.ac.uk)
19 The NHS Constitution for England - GOV.UK (www.gov.uk)
20 Paralympics history – Evolution of the Paralympic Movement
21 Disability since 1945 | Historic England
22 Disability history glossary | Historic England
23 The social model of disability – Sense
24 BBC – Four decades since Alf Morris' Landmark Disability Act
25 The Disability Discrimination Act 1995: The Campaign for Civil Rights – YouTube
26 https://www.equalityhumanrights.com/sites/default/files/being-disabled-in-britain.pdf
27 Building for equality: Disability and the built environment (parliament.uk)
28 Paralympics history – Evolution of the paralympic movement
29 2012 Paralympics ends, breaks all the records – CBS News
30 Locog hails biggest and best paralympics in history | Paralympics 2012 | The Guardian
31 A year after the Paralympics attitudes to disability need to improve | Paralympics 2012 | The Guardian
32 The public's perception of disabled people needs to change — we're not just Paralympians or scroungers | The Independent | The Independent
33 Disability Perception Gap | Disability charity Scope UK
34 'Transformation' in British attitudes towards disabled people since Paralympics 2012 – GOV. UK (www.gov.uk)

35 Prince Harry launches Invictus Games for injured troops at Olympic Park | Invictus Games | The Guardian
36 Braye, S., Gibbons, T. and Dixon, K. (2013). Disability 'Rights' or 'Wrongs'? The Claims of the International Paralympic Committee, the London 2012 Paralympics and Disability Rights in the UK. *Sociological Research Online*, 18(3), pp. 1–5.
37 Social model of disability | Disability charity Scope UK
38 Disabled adults four times worse off financially than non-disabled adults finds new DBC report | Disability Rights UK
39 jrf_-_uk_poverty_2019-20_findings_1.pdf

References

BBC. (2010). *Four decades since Alf Morris' landmark disability act.* [online] Available at: http://news.bbc.co.uk/local/manchester/hi/people_and_places/newsid_8697000/8697567.stm#:~:text=The%20Act%20has%20been%20described,support%20for%20people%20with%20disabilities [Accessed 14 January 2024].

BBC. (2022). *Arthur Hughes: First disabled Richard III is 'big gesture' from RSC.* [online] Available at: https://www.bbc.co.uk/news/entertainment-arts-61549419 [Accessed 14 January 2024].

BBC. (2022). *Census 1921 – 100-year-old secrets revealed.* [online] Available at: https://www.bbc.co.uk/news/uk-59879470 [Accessed 14 January 2024].

Blunkett, D. (2015). *The public's perception of disabled people needs to change — We're not just Paralympians or scroungers.* The Independent. [online] Available at: https://www.independent.co.uk/voices/comment/the-publics-perception-of-disabled-people-needs-to-change-were-not-just-paralympians-or-scroungers-10104704.html [Accessed 14 January 2024].

Branson, J. and Miller, D. (2002). *Damned for their difference: The cultural construction of deaf people as disabled.* Gallaudet University Press. p. 7.

Braye, S., Gibbons, T. and Dixon, K. (2013). Disability 'Rights' or 'Wrongs'? The Claims of the International Paralympic Committee, the London 2012 Paralympics and Disability Rights in the UK. *Sociological Research Online*, 18(3), pp. 1–5.

CBS News. (2012). *2012 Paralympics ends, breaks all the records.* [online] Available at: https://www.cbsnews.com/news/2012-paralympics-ends-breaks-all-the-records/ [Accessed 14 January 2024].

Christiansen, S. (2013). *A year after the Paralympics attitudes to disability need to improve.* [online] Available at: https://www.theguardian.com/sport/blog/2013/aug/24/paralympics-sophie-christiansen-equestrian [Accessed 14 January 2024].

Cohen, D. (2001). *The war come home.* Berkeley: University of California Press.

Disability Rights UK. (2019). *Disabled adults four times worse off financially than non-disabled adults finds new DBC report.* [online] Available at: https://www.disabilityrightsuk.org/news/2019/july/disabled-adults-four-times-worse-financially-non-disabled-adults-finds-new-dbc-report [Accessed 14 January 2024].

Dixon, Smith and Touchet. (2018). *Disability perception gap – Policy report.* Scope. [online] Available at: https://www.scope.org.uk/campaigns/disability-perception-gap/ [Accessed 14 January 2024].

Equality and Human Rights Commission. (2017). *Being disabled in Britain: A journey less equal.* [online] Available at: https://www.equalityhumanrights.com/sites/default/files/being-disabled-in-britain.pdf [Accessed 14 January 2024].

Gov.uk. (2014). *'Transformation' in British attitudes towards disabled people since Paralympics 2012 – GOV.UK.* [online] Available at: https://www.gov.uk/government/news/transformation-in-british-attitudes-towards-disabled-people-since-paralympics-2012 [Accessed 14 January 2024].

Gov.UK. (2023). *The NHS Constitution for England.* [online] Available at: https://www.gov.uk/government/publications/the-nhs-constitution-for-england/the-nhs-constitution-for-england [Accessed 14 January 2024].

Hasted, R. (2016). Domestic housing for disabled veterans 1900–2014 introductions to heritage assets. Historic England.

Historic England. (n.d). *Disability in the medieval period 1050–1485.* [online] Available at: https://historicengland.org.uk/research/inclusive-heritage/disability-history/1050-1485/ [Accessed 14 January 2024].

Historic England. (n.d). *Disability from 1485–1660.* [online] Available at: https://historicengland.org.uk/research/inclusive-heritage/disability-history/1485-1660/ [Accessed 14 January 2024].

Historic England. (n.d). *The lives of people with disabilities in 18th century England.* [online] Available at: https://historicengland.org.uk/research/inclusive-heritage/disability-history/1660-1832/the-lives-of-people-with-disabilities/ [Accessed 14 January 2024].

Historic England. (n.d). *Disability in the 19th century.* [online] Available at: https://historicengland.org.uk/research/inclusive-heritage/disability-history/1832-1914/ [Accessed 14 January 2024].

Historic England. (n.d). *Disability in the early 20th century 1914–1945.* [online] Available at: https://historicengland.org.uk/research/inclusive-heritage/disability-history/1914-1945/ [Accessed 14 January 2024].

Historic England. (n.d). *The impact of the First World War.* [online] Available at: https://historicengland.org.uk/research/inclusive-heritage/disability-history/1914-1945/war/ [Accessed 14 January 2024].

Historic England. (n.d). *Disability since 1945.* [online] Available at: https://historicengland.org.uk/research/inclusive-heritage/disability-history/1945-to-the-present-day/ [Accessed 14 January 2024].

Historic England. (n.d). *Disability history glossary.* [online] Available at: https://historicengland.org.uk/research/inclusive-heritage/disability-history/about-the-project/glossary/#cat_S_word_Social%20model [Accessed 14 January 2024].

House of Commons Women and Equalities Committee. (2017). *Building for equality: Disability and the built environment.* [online] Available at: https://publications.parliament.uk/pa/cm201617/cmselect/cmwomeq/631/631.pdf [Accessed 14 January 2024].

International Paralympic Committee. (n.d). *Paralympics History.* [online] Available at: https://www.paralympic.org/ipc/history#:~:text=On%2029%20July%201948%2C%20the,who%20took%20part%20in%20archery [Accessed 14 January 2024].

Joseph Rowntree Foundation. (2020). *UK Poverty 2019/20.* [online] Available at: https://www.jrf.org.uk/sites/default/files/migrated/files-research/jrf_-_uk_poverty_2019-20_findings_1.pdf [Accessed 14 January 2024].

Mawson, T.   (1918). *An imperial obligation: Industrial villages for partially disabled soldiers & sailors.* London: G. Richards, pp. 45–49.

Mercer and Barnes. *Chapter 1 – Changing disability policies in Britain – (From Barnes, C and Mercer G. (eds.) 2004: Disability Policy and Practice: Applying the Social Model, Leeds: The Disability Press, pp. 1–17).* [online] Available at: https://disability-studies.leeds.ac.uk/wp-content/uploads/sites/40/library/Barnes-policy-and-practice-1.pdf [Accessed 14 January 2024].

Rex, R. (2006). *Henry VIII and the English Reformation.* 2nd ed. Basingstoke: Palgrave Macmillan, p. 45.

Scope. (2015). *The Disability Discrimination Act 1995: The campaign for civil rights.* [online] Available at: https://www.youtube.com/watch?v=dwP1xuZZFuY [Accessed 14 January 2024].

Scope. (n.d). *Social model of disability.* [online] Available at: https://www.scope.org.uk/about-us/social-model-of-disability/ [Accessed 14 January 2024].

The Guardian. (2014). *Prince Harry launches Invictus Games for injured troops and Olympic Park.* [online] Available at: https://www.theguardian.com/uk-news/2014/mar/06/prince-harry-invictus-games-injured-troops-olympic-park [Accessed 14 January 2024].

The Historic England Blog. (2018). *Home from the war: What happened to disabled First World War veterans.* [online] Available at: https://heritagecalling.com/2018/12/14/home-from-the-war-what-happened-to-disabled-first-world-war-veterans/ [Accessed 14 January 2024].

Topping. (2012). *Locog hails biggest and best Paralympics in history.* [online] Available at: https://www.theguardian.com/sport/2012/sep/06/paralympics-ticket-sales [Accessed 14 January 2024].

University of Leicester. (n.d). *Richard III: Discovery and identification – Scoliosis.* [online] Available at: https://le.ac.uk/richard-iii/identification/osteology/scoliosis [Accessed 14 January 2024].

Chapter 3

Defining Disability

Introduction

It is recognised that a wide and diverse range of conditions may result in an individual being classed or categorised as disabled. Most of these conditions are either physical or mental, which are diagnosed by medical professionals. However, despite the notion that a medical diagnosis may engender clarity or reassurance concerning such conditions, there is a sense of individuality concerning the definition of disability. There is a combination of broad classifications concerning the groups of disability, and added to this are a number of umbrella terms used to group complex medical conditions. Irrespective of this, it is necessary to consider disability from the perspective of those individuals experiencing it. There is no doubt that there is a level or pain or discomfort suffered by many with disability, but it is necessary to additionally consider how those with disability interact with society. The ability to interact with society and undertake activities considered to be normal in daily life is a key feature required to define disability.

Disability has no boundaries in the sense that it does not differentiate between the wealthy and the poor. However, those in poverty or in geographically underdeveloped locations no doubt has less support or opportunity for social integration. This further Emphasises the important part society plays in defining disability. Having a physical or mental condition is widespread throughout the world, and accordingly, there are established universal definitions.

Global Recognition

Recognising the various attempts to define disability, the Council of Europe concludes that there is no universal definition, but it does explicitly refer to that definition adopted by the World Health Organisation (WHO). This indicates that disabilities are impairments limiting activity and participation. The WHO definition of disability goes further explaining that this as *a complex phenomenon, reflecting the interaction between features of a person's body and features of the society in which he or she lives*. This definition of disability also identifies the significance of environmental and social barriers within society as a contributing factor.

Disability is a global issue, and accordingly, the United Nations (UN) has adopted a definition within the Convention of Human Rights, Article 1 (2006) as *those who have long term physical, mental, intellectual or sensory impairments*. The definition goes on to

DOI: 10.1201/9781003239901-3

recognise that disability prevents *normal day-to day activities* and *may hinder their full and effective participation in society on an equal basis with others.* The use of the word 'equal' in the context of service provision is a strong standpoint when compared to the stance of the Equality Act 2010 in the UK, which obliges a 'reasonable adjustment' to facilitate inclusivity.

In line with the European Council's assertion that there is no universal definition of disability, the European Union (EU) explicitly refers to case law determined by the Court of Justice. Accordingly, this definition includes an impairment that is *long-term* and which, in the field of professional life, *hinders an individual's access to, participation in, or advancement in employment.*

UK Definition

The specific legislation associated with the provision of disabled access to goods, services and employment in the UK will be discussed thoroughly in Chapter 4. However, by means of an introduction to the key principle of definitions in the UK, The Equality Act 2010 defines disability as either a *physical or mental impairment* which must have a *substantial long-term effect* on a person's ability to carry out *normal day-to day activities.* This Act was not the first attempt to formalise the definition of disability as The Disability Discrimination Act 1995 (DDA95) was the first piece of bespoke legislation aimed specifically at the provision of inclusive environments. The DDA 1995 was revised in 2005 and then repealed in 2010 to make way for the Equality Act, which encompasses the wider coverage of a series of protected characteristics including disability. The definition of disability under the Equality Act 2010 has been adopted for the purpose of this research, although it should be noted that this effectively is a medical definition of disability as opposed to a social model or definition which is discussed later in this chapter.

Universally, the definitions of disability recognise that this is something that is not temporary but long term and ultimately limits, restricts or prohibits participation in normal day-to-day activities. It is necessary to contextualise this into the different disabilities and how this applies to building owners, service providers and users. Current attitudes to disability and inclusion are perceived to have evolved over the past 50 years.[1] However, in reality, there are many historical references to disability in the UK for hundreds of years, and it is interesting to see how attitudes appear to have evolved from the early benevolent approach of society to current assertive attitudes of individuals regarding equal rights. This has been extensively discussed in Chapter 2.

Social Model

As previously alluded to, there is a need to consider the social interaction in the definition of disability, and accordingly, the social model of disability addresses this topic. Coined in 1983 by Mike Oliver, the term *social model of disability* suggests that it is not the person with a physical or mental impairment who is disabled, rather society that makes them disabled.[2] This idea is exemplified within the built environment by entrance steps to a building that create disability and a ramp that may remove the disability. Further examples may include the presence of audio description for the blind or subtitles for the deaf. All of these examples under the definition of the social model of disability place an emphasis on society to remove disability. Breaking down these barriers appears more straightforward in theory than in practice. As with the wide variety of disabilities and indeed the nuances within grouped disabilities or those with umbrella terms, the solutions provided by society can vary in complexity.

It is recognised that there are probably 200 years of commercial development in society by which there have been those trading goods or delivering services to the general public. As established in the discussion surrounding the history of disability in the UK, while there have been pockets of society supporting those with disability, the formal requirement to design for disability has only been evident since the very latter part of the 20th century. Looking at commercial properties built in the past 200 years, close to 90% of these have not had to consider access. As a consequence, they contribute significantly to the social model of disability. It is therefore necessary to consider the characteristics of commercial property and the complexity to retrofit this property for accessibility; this is discussed fully in Chapter 5. However, prior to this, it is necessary to assess the demographics of disability in the UK to ascertain specific characteristics of both physical and mental impairments. Only then can the individual disability-specific needs be established to deliver solutions to address the social model of disability.

Demographics

According to data held by the UK government and published in the public domain,[3] it is estimated that more than one in five people in the UK reported a disability in 2020/21, which amounts to 14.6 million people and equates to 22% of the population (Gov.UK, 2023) (Fig 3.1).

This is an overall increase of 26% from 2010/11 (11.6 million), although significantly, the percentage of state aged person who recognise as having a disability has dropped from 45% to 42% from 2019/20 to 2020/21. This is significant as this may reflect the above average numbers of recorded deaths attributable to the COVID-19 pandemic. Such basic data aligns with the BBC reporting that 60% of all deaths attributable to the pandemic occurred in those with disability.[4] The prevalence of disability amongst the elderly and also the high death rate associated with this age group is therefore the likely reason that there was an overall drop in the percentage of those recognising as having a disability in 2020/21.

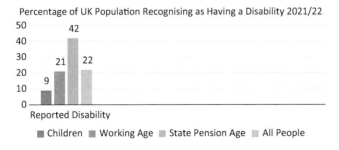

Figure 3.1 Percentage of UK population reporting disability 2020/21 (Gov.UK, 2023).

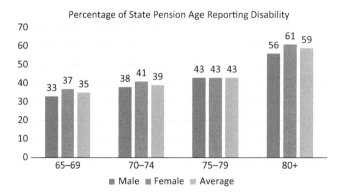

Figure 3.2 **Percentage of state pension age reporting disability (Gov.UK, 2023).**

Looking further at the government data within the state pension age; 61% of people above the age of 80 report a disability, while of those aged 65–69; 35% reported disability (Fig 3.2):

Concerning the types of reported disabilities; mobility at 51% is the largest and compared to 2014/5, mental health has increased by 20%. According to the survey data the percentages of reported disabilities has remained largely the same since 2014/5 (Fig 3.3):

Considering the overall percentage of the UK population reporting disability is 22%, it is necessary to analyse any regional variations. According to Gov.UK (2023), the regions reporting the highest percentage of disability are the Northeast (31%) and Wales (28%). The region reporting the lowest percentage of disability is London (15%) (Fig 3.4):

The notion that there is less incidence of disability in London and the Southeast is probably due to a young working-age demographic who statistically are less likely to recognise as being disabled. The obvious key fact to draw upon and, as previously discussed, is the reported incidence of disability that increases with age and particularly above the state pension age. Concerning those with disability, it has been noted the

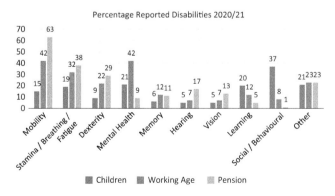

Figure 3.3 **Percentages of reported disabilities 2020/21 (Gov.UK, 2023).**[5]

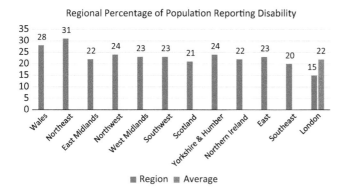

Figure 3.4 Regional percentage of population reporting disability (Gov.UK, 2023).

notion of increased levels of poverty amongst those with disability, and it was also necessary to consider the prevalence of **disability and poverty** in the older population.[6] These two factors are important when considering access to commercial goods, services and employment since, while it is important to open up commercial opportunity to those with disability, there will be a number of this demographic who do not have the financial means, irrespective of access.

While those aged 65 and above have a higher incidence of disability, within the working-age category; those reporting disabled are just under 1 in 5. This is important regarding the provision of inclusivity in the workplace as also obliged under the Equality Act 2010. Disabled people make great employees and can contribute to the workplace; however, a number of barriers or challenges may exist. While there is a requirement to make reasonable adjustment to access the workplace, there is also a need for employers to recognise the needs of disability in their workforce and make operational adjustments.

In conclusion, a national average relating to 22% disability of the population would appear to suggest that this may lead to a positive attitude to the need for inclusive provision and a resonance with the overall positivity experienced with high-profile events such as the Paralympics. Despite this, it is necessary to analyse further both physical and mental impairments that shape the definition of disability. It may be argued that society's view of disability may be informed by those with visible conditions affecting mobility or vision, and this is often exemplified by the extraordinary actions of those participating in the Paralympics. The reality of government data is such that those with disability often have more than one classification of condition and approximately one in five have other, often unseen disabilities. Compounded to this is the high percentage of those of pension age who have a significantly higher incidence or recognition of long-term physical or mental impairment that prevents participation in normal day-to-day activities. This demographic in society has probably contributed in the main the most to society. They may also be considered the 'forgotten' generation in the context of disability. Society appears less inclined to view those above the age of 65 with physical or mental impairment to be 'disabled'. There appears a notion that this sector of society is 'elderly' and as such has conditions associated with 'old age'. Under the terms of the Equality Act 2010, this age demographic is protected against discrimination through both age and disability qualifying as 'protected characteristics'.

Considering that disability crosses so many geographic and societal frontiers, it is necessary to consider the concept and definition of some common physical and mental impairments.

Physical and Mental Impairments

To begin to understand the access requirement to commercial property or barriers to employment, it is necessary to examine a range of disabilities. It has been established that age is an important factor in the contribution of those who consider themselves to have a long-term physical or mental impairment, with those aged 65 or above more likely to have mobility or stamina issues. Another example of an age-related disability concerns mental health, which appears more preventable amongst those of working age. Furthermore, social or behavioural issues appear more widespread in children. One important consideration when auditing and implementing accessible design is to note the term **ambulant disabled**. These are individuals who recognise as having a physical or mental impairment but do not require the regular use of a wheelchair. This term does cover a wide range of disabilities, and it is necessary to contextualise it within the categories of disability used by the UK government; these are divided into the following sub-groups:

- mobility;
- stamina and breathing;
- dexterity;
- mental health;
- memory;
- hearing;
- vision;
- learning;
- social and behavioural and
- other (including hidden disabilities).

The data indicates that those who recognise as having a physical or mental impairment often have more than one condition. This again may be more evident with the elderly, who could struggle with mobility and stamina or breathing as well as vision or memory loss.

The individual types of disabilities or medical conditions are varied, and in many cases, complicated. There are often a range of different symptoms, and the inter-relationship between these can be complex. The complete opposite can be said regarding the pragmatic approach to designing for disability, and this is detailed fully in Chapter 9. In essence, access audit as discussed in Chapter 6, and the provision of features to promote an inclusive environment are black or white, buildings are either compliant or not. The framework for this is driven by extensive research to generate building regulations or design standards. This 'mechanical' approach aims to be objective and unambiguous but also results in 'one size' design, which is not only difficult to implement retrospectively to existing buildings but seeks to adopt a blanket approach to designing for disability. The following is a **non-medical analysis** of both physical and mental impairments. The information has been taken and summarised from publicly accessible recognised charities supporting these disabilities or information provided by The National Health Service (NHS). The range of disabilities has been contextualised with a summary of the evidenced-based research considering the needs of those in accessing buildings, goods, services or for employment. The general access requirements are graded 1 to 5 where 1 is not considered not problematic and 5 likely to prevent access to goods and services:

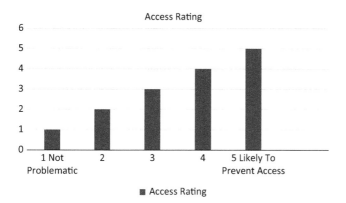

Access Rating

Figure 3.5 Grading of disability specific access requirements.

Fig 3.5

Alzheimer's and Dementia

A summary of the definition and information by Alzheimer's Disease International (www.alzint.com) is detailed below:

Often associated as a condition affecting those over the age of 65, dementia can occur in younger people, and this is known as young onset dementia. The term 'dementia' is used to describe a collection of conditions, with **Alzheimer's being the most common** and accounting for about half of the diagnoses.

Presently dementia is a condition that is not curable and causes damage to the nerves in the brain with a range of symptoms, including:

- memory loss;
- inability to find the right words or comprehend what others are saying;
- difficulty in undertaking previously routine tasks and
- changes in mood and personality.

Dementia suffers are unable to care for themselves and accordingly require help or assistance with many facets of daily life. The condition has high prevalence in old age and is an example of a disability that many will experience directly or indirectly in the course of their lives.

While dementia does not typically result in a physical impairment affecting mobility, there is an increased incidence of this condition with the over 65, and as a consequence, there is also a

(Continued)

likelihood of physical impairment. When considering the access issues affecting those who categorise their physical or mental impairment to be mobility, the following data has been collected:

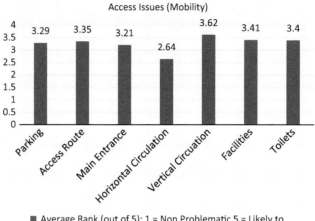

Access Issues (Mobility)

■ Average Rank (out of 5): 1 = Non Problematic 5 = Likely to Prevent Access

Amputee

Summary of definition and information provided by the National Health Service [NHS] (www.nhs.uk/conditions/amputation/).

Amputation typically concerns the full or partial removal of an arm or leg and is usually done for one of the following reasons:

- severe infection;
- a limb is affected by gangrene;
- complications of diabetes;
- serious trauma from a crush or blast wound or
- a limb is deformed and has limited movement or function.

Post amputation, there is a need for physical rehabilitation, and it is recognised that this can be a long, difficult and frustrating process. There is the possibility that a prosthetic limb can be fitted, and learning to use it can form part of the rehabilitation. Several clinical decisions influence the fitting of a prosthetic limb, and is it important to note that learning to use this alongside the support of a physiotherapist is a difficult and demanding process.

In terms of accessing commercial properties, goods, services and employment, amputees will have differing disability-specific needs depending on what limb has been amputated. It primarily is likely to affect mobility but could also be classed within the categories of stamina and breathing as well as dexterity.

There were insufficient responses received in the research data concerning the disability-specific access requirements of those recognising as being amputees. In line with some of the general information available for this disability, there was noted a high prevalence in the use of wheelchairs and walking aids. Furthermore, there is a recognition within the limited responses that alongside mobility, strength, stamina and dexterity are also appropriate classifications for this physical impairment.

Arthrogryposis Multiplex Congenita (AMC)

A summary of the definition and information by the Arthrogryposis Group (www.arthrogryposis.co.uk) is detailed below:

Arthrogryposis Multiplex Congenita (AMC) is a term used to describe over 300 conditions that cause multiple curved joints in areas of the body at birth. The severity of the condition is varied between individuals, and the condition is non-progressive. While there are many variants associated with this condition, the most common can be placed into the following categories:

- disorders with mainly limbs;
- disorders with limb plus some other body area, e.g., cleft lip, the heart, intestinal defects, curvature of the spine and
- disorders with limb and the central nervous system.

AMC is not considered to be a genetic condition, and some of the specific nuanced conditions respond well to physical treatment upon initial diagnosis. The research did not capture sufficient responses for this condition, but it can be summarised as a physical impairment likely to be classified to affect mobility, and accordingly, the research data for this category of impairment is detailed below:

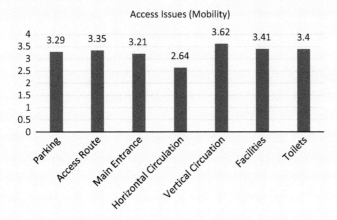

Access Issues (Mobility)

■ Average Rank (out of 5): 1 = Non Problematic 5 = Likely to Prevent Access

Attention Deficit Hyperactivity Disorder (ADHD)

A summary of the definition and information by the National Health Services [NHS)] (www.nhs.uk/conditions/attention-deficit-hyperactivity-disorder-adhd/) is detailed below:

ADHD is a condition that can affect both children and adults; it is associated with difficulties in concentrating as well as affecting behaviour. People with ADHD may appear restless or present impulsive behaviour. Typically, it may become apparent at school in the learning environment, but sometimes there is later life diagnosis.

Attention deficit hyperactivity disorder (ADHD) is a condition that affects people's behaviour. People with ADHD can seem restless, may have trouble concentrating and may act on impulse. Those with ADHD may also have anxiety or sleep-related problems, and some of the challenges might include organisation of tasks or being able to follow specific instructions. As a consequence of ADHD, time management may be difficult, and the inability to focus or complete tasks may be evident. This could lead to a sense of restlessness or impatience, which can engender an impulsive or risk-taking behaviour.

(Continued)

People with ADHD may also have difficulty coping with stress, stressful situations or social interactions.

It is recognised that ADHD is a neurological condition that is more prevalent in boys than girls and persists into adult life. There were minimal survey responses specifically citing ADHD as a disability, and this appeared prevalent to a small number of respondents as part of other more complex physical and mental impairments. As a standalone disability, ADHD does not appear to have any specific challenges when accessing commercial buildings, goods and services. There is no doubt the appearance of condition-specific requirements concerning employment, but no trends concerning this or issues accessing the built environment could be established with so little primary data.

Ankylosing Spondylitis/Axial Spondyloarthritis (SpA)

A summary of the definition and information by the National Axial Spondyloarthritis Society (www. nass.co.uk) is detailed below:

Axial spondyloarthritis is often abbreviated to Axial SpA and is a term used to define a number of conditions accounting for **changes to joints in the spine, lower back and pelvis**. The symptoms include:

- a gradual development of back pain over a number of weeks or months;
- back pain typically evident at the start of the day and reducing with activity or movement throughout the day but returning with rest;
- a long-term condition as opposed to broken periods of back pain;
- possible weight loss, fever and night sweats;
- the symptoms can vary in intensity or change from day to day and
- can come in periods of relative sustained discomfort known as flares.

Inflammation can occur or be evident to other parts of the body including Uveitis to the eyes, inflammatory bowel disease, tendon inflammation to tendons or ligaments attaching to bones but often to the ankle or foot. Axial SpA can also cause skin inflammation.

The research data suggests that there is little correlation between age and the condition, and while the majority of respondents recognise that this is a mobility issue, many consider this to be in part a hidden disability, which may be the case with inflammatory bowel disease.

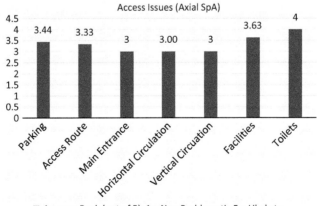

Access Issues (Axial SpA)

■ Average Rank (out of 5): 1 = Non Problematic 5 = Likely to Prevent Access

Arthritis

A summary of the definition by Arthritis Action (www.arthritisaction.org.uk) is detailed below:

Arthritis is a condition that results in the inflammation or swelling of joints in the body. The most common type is osteoarthritis, which is often associated with old age. Rheumatoid arthritis occurs when the immune system attacks the joints causing inflammation.

The typical symptoms of arthritis are:

* swollen joints;
* pain and discomfort to joints and
* stiffness and restrictive movement.

There are a number of causes of arthritis, and this can be a genetic condition as well as gender-related for some conditions where women are more affected than men. Some conditions are age related, with osteoarthritis typically occurring later in life, or it can be linked to a historic injury or damage to a joint. In some instances, arthritis can be linked to lifestyle with alcohol or smoking contributing to specific conditions along with being overweight. All of these causes are a contributing factor to the high incidence of this physical impairment throughout the population.

The research data has identified that those with arthritis categorise this mostly as a physical impairment affecting mobility and dexterity. Accordingly, the data has also established most of those respondents with arthritis are ambulant disabled and not wheelchair users. The approach to accessing buildings, goods, and services is challenging with vertical circulation ranking the most problematic.

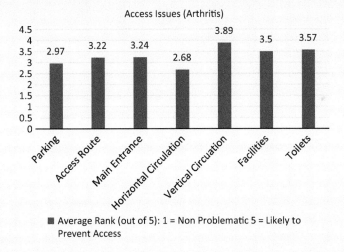

Access Issues (Arthritis)

Average Rank (out of 5): 1 = Non Problematic 5 = Likely to Prevent Access

Autistic Spectrum

A summary of the definition and information by the National Autistic Society (www.autism.org.uk) is detailed below:

Autism or the autistic spectrum is an umbrella term covering a wide variety of symptoms affecting both children and adults. Autistic people share several possible difficulties, such as social communication and literal understanding. They may find it difficult to interact with others and appear withdrawn. In contrast, some of their expressive behaviour might be considered inappropriate, and they might exercise

(Continued)

repetition in aspects of behaviour or routine. Changes in routine can be confusing or lead to anxiety, while they may also have differing reactions to the senses of taste, touch, sound or light levels.

Autistic people may have intense focused interests or hobbies, and this is recognised as contributing to individual personal enjoyment or wellbeing. Another characteristic of people on the autism spectrum is levels of anxiety, which can be extreme in some cases. This might be evident in social situations or when faced with individual challenges. The autistic spectrum is recognised as being varied, and it is important not to assume all those with autism express this the same way. It should be further acknowledged that people with autism can appear to shut down or melt down when faced with overwhelming situations. This can appear as outwardly expressing behaviour or appearing withdrawn and quiet.

The research data identified a number of respondents who also recognised as having some physical impairments as well as some reporting mental impairments such as mental health, ADHD and Specific Learning Disabilities (SpLD) such as dyslexia and dyspraxia. When analysing the data from a limited number of respondents, a trend suggests that access of commercial buildings, goods and services is less of a barrier; those experiencing access are likely to be affected by other physical impairments.

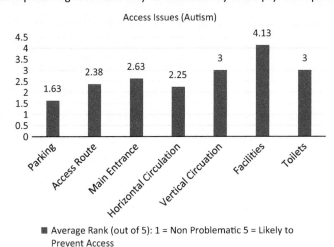

Bechet's Disease

A summary of the definition and information by Bechet's UK (www.behcetsuk.org) is detailed below:

Behçet's Syndrome (also known as Behçet's Disease or, simply, Behçet's) is a chronic condition relating to the body's immune system, which normally protects against infection. An over-reactive immune system typically causes inflammation of the blood vessels, arteries and veins in the body. The condition is considered painful but not life threatening, although it is considered rare. There is no cure for Bechet's, although medical intervention can help manage the condition, which can include suppression of the immune system. There is also a wide range of symptoms and these can affect the following operations of the human body:

- neurological;
- central nervous system;
- eyes;
- cardiovascular system;
- chronic fatigue;
- joints;

(Continued)

- extremities;
- skin;
- genital area;
- gastro-intestinal;
- mouth and
- psychological.

With such a wide range of symptoms, it is anticipated that there would be a range of issues and requirements across the key access criteria. However, there were no respondents to the research survey, and accordingly, it is not possible to ascertain the disability-specific access concerns. Contrasted against this assumption is confirmation by Bechet's UK that with appropriate medical help, those with Bechet's can lead close-to-normal daily lives.

Bell's Palsy

A summary of the definition and information by Bell's Palsy (www.bellspalsy.org.uk) is detailed below:

Bell's Palsy is facial paralysis, resulting from a dysfunction of the 7th cranial nerve. The condition normally affects one side of the face and can last from several weeks to several years in extreme cases. The majority of instances reduce in severity after a few months; however, there are some in which the symptoms last indefinitely. The symptoms include:

- drooping or sagging of the mouth;
- facial pain;
- drooling;
- loss of taste and
- inability to close the eye resulting in tearing.

Those with Bell's Palsy may need to take time off work as it is recognised that recovering from the condition can be very tiring. When placed into the context of the built environment and accessing commercial property, goods, services and employment, the symptoms do not appear to relate to those likely to present a barrier to access.

There were no survey responses from those with Bell's Palsy, and the nature of this being a 'temporary' condition for about 80% of those diagnosed may be a contributing factor in the lack of response. Furthermore, and while it is accepted that for 20% of those diagnosed with Bell's Palsy, it can be a long-term condition, the interpretation of the Equality Act 2010 defines disability as being long term.

Blindness and Visual Impairment

A summary of the definition and information by the Royal National Institute of Blind People (www.rnib.org.uk)

Blindness and visual impairment are an umbrella term used to describe people with sight loss. There are a variety of conditions causing sight loss, and these include:

- age-related macular degeneration (AMD);
- cataracts;
- diabetic retinopathy;
- glaucoma;

(Continued)

- uncorrected refractive error and
- other eye problems.

While some people with sight loss or blindness are born with these conditions, many other symptoms of sigh loss is age related. While there are instances of intervention or corrective measures to address sight loss, some conditions lead to permanent blindness.

It is important to note that there is a difference between sight loss, which is often referred to as partially sighted, and severe sight loss or blindness. The research noted that a number of respondents recognising as having sight loss or severe sight loss also reported other 'mobility' conditions, and it is accepted that this may skew the data. In summary, this classification of disability recognises issues with accessing commercial buildings, goods, services and employment across the spectrum of accessible features.

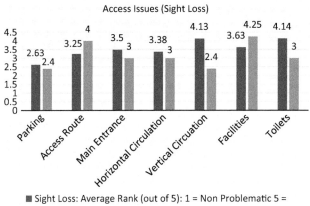

Access Issues (Sight Loss)

■ Sight Loss: Average Rank (out of 5): 1 = Non Problematic 5 = Likely to Prevent Access

■ Severe Sight Loss: Average Rank (out of 5): 1 = Non Problematic 5 = Likely to Prevent Access 2

Brain and Head Injury

A summary of the definition and information by Headway: (www.headway.org.uk)

The term brain or head injury covers a wide and complex variety of conditions, but in essence, these can relate to traumatic brain injury, typically caused by road traffic accidents, falls or from being assaulted. Less serious are minor head injuries or concussion, which could be attributed to less traumatic circumstances or sustained while undertaking sports such as rugby. An aneurysm is a condition in which there is a swelling in the weakened wall of a blood vessel in the brain; this can in turn lead to a brain haemorrhage if the swollen blood vessel ruptures. Brain tumours are another complex condition; these are growths within the skull and are either malignant or benign.

Carbon monoxide poisoning can cause brain injury, along with hypoxic and anoxic brain injury, resulting from a lack of oxygen to the brain; accordingly, there can be varying degrees of severity associated with it. Encephalitis is the swelling of the brain associated with infection, while meningitis is an example of a bacterial or fungal infection that can also cause swelling of the brain.

(Continued)

A blood clot in the vessels supplying the brain or rupture of these is a stroke, and this is considered an emergency condition.

Brain injuries can cause many effects such as:

* memory loss;
* changes in behaviour;
* difficulty in communication;
* physical effects or mobility challenges and
* tiredness.

Insufficient responses were received concerning this disability, although the wide spectrum of symptoms suggest that there may be many challenges with accessing commercial buildings, goods, services and employment. Another consideration is the severity of brain or head injury, which also contributes to access difficulties; therefore, making general statements on access that do not reflect the nuances of this condition should be avoided.

Brittle Bone Disease (Osteogenesis Imperfecta)

A summary of the definition and information by the Brittle Bone Society (www.brittlebone.org)

Osteogenesis Imperfecta (OI), which is also known as brittle bone disease, is a lifelong genetic condition that results in bones that are easily fractured or broken. It is a relatively rare condition, and bones are sufficiently fragile to break without significant impact or trauma. OI has varying degrees of severity, and those with the condition may require the use of mobility aids, such as a wheelchair, walking frames or walking sticks.

This is predominantly a physical disability that can also affect joints or present as bone curvature, brittle teeth and hearing loss. While there were no responses in the research survey data concerning this disability, it is necessary to consider the effects on mobility. The brittle bone society has identified the need for some of those with the condition to use wheelchairs, walking sticks and walking frames. Accordingly, this aligns with the general response concerning the mobility category

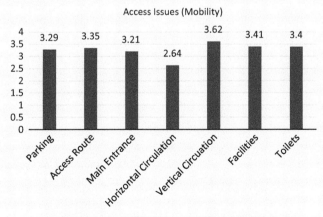

Access Issues (Mobility)

■ Average Rank (out of 5): 1 = Non Problematic 5 = Likely to Prevent Access

(Continued)

Cancer

A summary of the definition and information by the National Health Service [NHS] (www.nhs.uk/conditions/cancer/) and Cancer Research UK (www.cancerresearchuk.org):

Cancer is a term to cover a medical condition in which cells within the body reproduce of grow in an uncontrolled manner. As a consequence, this damages previously healthy tissue or organs. With over 200 known cancers, this condition affects 1 in 2 people during a lifetime; the most prevalent cancers are breast, lung, prostate and bowel.

The symptoms of cancer vary with the different types, and not all present as a physical or mental impairment in the terms of disability. There is a wide variety of condition-specific treatments that can contribute to impairment affecting normal day-to-day activities.

As cancer is a general term that covers a wide variety of different and complex conditions; therefore, access to commercial property, goods, services and employment will depend on the relevant sub-category, which could typically be mobility, stamina/breathing or dexterity. These examples of physical symptoms may not be present upon initial diagnosis but occur and become progressively worse with time.

The survey responses in the data collection were insufficient regarding the disability-specific requirements of those with cancer. One common theme for the limited, collective group of respondents was that those with cancer also identified several other different disabilities. This emphasises that cancer is a disease that can affect anyone, irrespective of age or gender, although some cancers are more gender specific.

Concerning the access challenges and the disability-specific requirement for accessing commercial properties, goods, services and employment, this is dependent on the type and stage of cancer. Accordingly, those who require the use of a wheelchair during the treatment or recovery from this disease may be classed as having a disability in the mobility category.

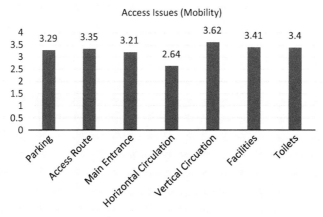

Access Issues (Mobility)

■ Average Rank (out of 5): 1 = Non Problematic 5 = Likely to Prevent Access

(Continued)

Cerebellar Ataxia

Summary of the definition and information by Ataxia UK (www.ataxia.org.uk):

Ataxia is a rare condition that can affect balance and coordination. This diagnosis can be temporary or permanent, with most cases being progressive. Currently, there is no cure for ataxia, and it is recognised that there are different types of ataxias. It can be a genetic condition or something that happens as a result of a brain or head injury. As the condition affects balance, people with this condition might at first note difficulty in walking in a straight line or have a sense of clumsiness. Other symptoms may include:

- slurred speech;
- difficulty in swallowing;
- tremors or shaking;
- a sense of tiredness and
- vision problems.

There were insufficient responses to the survey data request in the research concerning the disability-specific requirements for those with ataxia. It is anticipated that while the physical symptoms are likely to contribute to mobility, breathing, stamina and dexterity-related access requirements, it is also necessary to recognise the consequence of visual impairment.

Cerebral Palsy

Summary of the definition and information by Cerebral Palsy UK (www.cerebralpalsy.org.uk):

Cerebral Palsy is a term used to cover several neurological conditions and is caused by brain injury occurring before, during or shortly after birth. A bleed in the brain of a baby or problems with oxygen supply to the brain are causes, along with a premature birth or complications during the birthing process. It can also arise through the mother catching an infection during pregnancy or changes in the genes, which can affect development of the brain.

The symptoms of cerebral palsy can vary but typically these present as issues with:

- muscle control;
- coordination;
- reflexes;
- posture and
- balance.

There are a number of recognised variances in cerebral palsy, which also present as muscle stiffness or weakness (Spastic Cerebral Palsy), muscle tone causing involuntary spasms (Athetoid Cerebral Palsy) and problems with balance or coordination (Ataxia Cerebral Palsy).

Insufficient survey responses were received to ascertain conclusively the disability-specific requirements of those with Cerebral Palsy. However, this condition is recognised as being within the mobility and dexterity categories of disability, which present very similar access challenges.

(Continued)

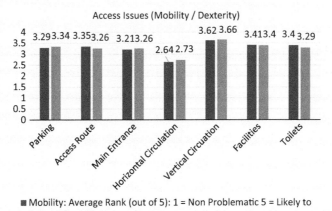

Access Issues (Mobility / Dexterity)

■ Mobility: Average Rank (out of 5): 1 = Non Problematic 5 = Likely to Prevent Access
■ Dexterity: Average Rank (out of 5): 1 = Non Problematic 5 = Likely to Prevent Access 2

Charcot Marie Tooth (CMT) Disease

Summary of the definition and information by Charcot-Marie-Tooth UK (www.cmt.org.uk):

Charcot-Marie-Tooth Disease is a genetic condition that results in damages to the peripheral nerves, which are used for muscle control. CMT also affects the sensory nerves, which transmit information to the brain concerning pain, heat, cold, touch and balance.

As a consequence, those with CMT have progressive muscle weakness, which is often evident to the hands and feet that can give a numbing sensation. It is a condition that can present different challenges as these progress, which can mean that those with the condition may have to adapt to these changes.

The findings of the research identify that this is a condition that is primarily classed as a mobility issue, but this also affects dexterity in about half of those with this condition. Noted also is the requirement of those with CMT to use a variety of mobility aids, with walking sticks being the most prevalent. This aligns with a number of specific challenges when accessing commercial properties, goods, services and employment.

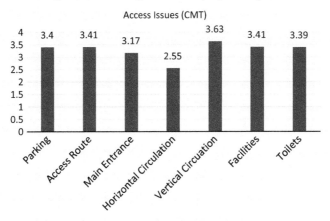

Access Issues (CMT)

■ Average Rank (out of 5): 1 = Non Problematic 5 = Likely to Prevent Access

Chronic Fatigued Syndrome (M.E.)

Summary of definition and information by Action for M.E. (www.actionforme.org.uk):

Myalgic Encephalomyelitis (M.E.) is recognised by medical professionals under the term chronic fatigue syndrome (CFS or CFS/M.E.). The symptoms of tiredness or fatigue are extreme and far from the tiredness that might be felt after a poor night of sleep.

Those with the condition struggle to get started in the day and can also have post-exertion tiredness, which is seen by some and 'pay back' for physical activity. The symptoms of M.E. are very wide ranging, and not all those with the condition present with the same combination. These can include:

- an inability to recover from small amounts of exertion;
- pain in muscles or joints with pins and needles;
- muscle twitches or cramps;
- headaches or migraine;
- stomach complaints and
- a number of sleep-related difficulties.

The symptoms are so varied that they can also include flu-like symptoms, bowel irritation through to 'brain fog' and dizziness. A limited number of respondents were noted in the research, and accordingly, it is difficult to confirm completely the trends in disability-specific access requirements. However, it was noted that respondents were less likely to access commercial properties goods or services every day, and this may be as a direct result of chronic fatigue. Alongside a recognition that those with this disability classify this as a mobility issue, there is also a recognition that this is classed as a stamina and breathing issue.

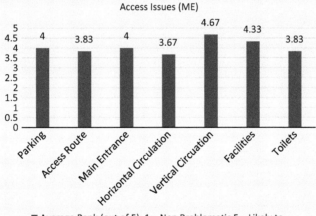

Access Issues (ME)

■ Average Rank (out of 5): 1 = Non Problematic 5 = Likely to Prevent Access

Cystic Fibrosis (CF)

Summary of the definition and information by the cystic fibrosis Trust (www.cysticfibrosis.org.uk):

Cystic fibrosis (CF) is a genetic condition which causes the body to produce thick mucus affecting the lungs and digestive system. It is considered a life-limiting condition and may not be obviously visible, the symptoms include:

- Poor lung function;
- frequent and persistent lung infections and
- the inability to effectively digest food, particularly fats.

(Continued)

Cystic fibrosis is not something that can be caught later in life as it is a condition that is present at birth. While everyone needs mucus in the lungs to trap particles and bacteria, CF leads to the build-up of stickier and thicker mucus which can lead in turn lead to lung infections and possible lung damage. Additional complications include CF-related diabetes, bone disease and infertility.

The treatment of cystic fibrosis can include physiotherapy to loosen the mucus which is known as airway clearance.

There were insufficient responses in the research data from those with cystic fibrosis therefore it is not possible to ascertain any disability specific requirements or the access of commercial properties, goods, services and employment. While it may be suggested that those with CF would fall mainly within the stamina and breathing category of disability, the condition also contributes to thinner and weaker bone development, this can contribute to joint pain and conditions such as osteoporosis. Cystic fibrosis can also result in inflammation and scarring to the pancreas which can cause a type of diabetes. Significant problems with the digestive system and also possible liver disease are other complications that can arise from the condition.

In summary cystic fibrosis is not as rare as many other medical conditions, it can have some significant complex implications although it is not always obvious when someone has this. Therefore, it can be said that CF is, for some, a hidden disability.

Deafblindness

Summary the definition and information by Deaf Blind UK (www.deafblinduk.org.uk):

Deafblindness is a term which describes the condition of having an impairment of both sight and hearing to varying degrees. It is a severe disability which presents problems with communication, mobility and accessing information. It is significantly more severe than those suffering either deafness or blindness alone as affecting both of these senses means there is little possibility of one of the other compensating for this.

There are a variety of reasons for deafblindness but these can be divided into those who have acquired this and this is the majority of instances. Either deafness or blindness may occur independently and progressively through a range of factors including old age, alternatively deafblindness is congenital.

The absence of any respondents in the research data has resulted in the need to rely upon the guidance provided by Deaf Blind UK in respect of difficulty accessing commercial property, goods, services and employment. The observation that many of those who recognise as being deafblind have acquired this as opposed to being born with the condition is influential concerning access.

The notion that someone born deaf or with sight loss can acquire deafblindness with age is logical. Added to this is evidence that those over the age of 65 are more likely to recognise as having mobility or breathing and stamina issues further reinforces that deafblind people also have problems with mobility. Accordingly, those who classify their disability as primarily affecting mobility identify vertical circulation as being the biggest contributing factor in preventing access.

In summary deafblind as a disability has a high level of needs to facilitate access to goods, services and employment.

Deafness and Hearing Loss

Summary of definition and information by the Royal National Institute for Deaf people (www.rnid.org.uk):

Hearing loss affects many million people in the UK and represents a large quantity of those who recognise as being disabled.

This may be gradual and is often associates with old age, typical observations of hearing loss are having to turn up the TV louder than say with other family members, struggling to follow conversations in public places such as pubs or restaurants or asking to people to repeat things that are said. There is also a sense that during conversation, others appear to be mumbling or partners complain that they are not being listen to.

Hearing loss can also be instant if someone is exposed to extremely high levels of noise or sound which causes permanent damage to the hearing components within the ear or in some cases as a consequence of other disabilities detailed in this text book.

Exposure to high levels of noise or sounds can also cause tinnitus which presents as

ringing, whistling or whooshing inside the head or ears. Tinnitus is often, but not always, linked to hearing loss.

The research has identified that as disability, deafness or hearing loss are less likely to have the same level of access concern as other physical or mental impairments.

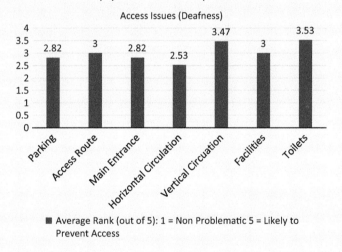

Access Issues (Deafness)

■ Average Rank (out of 5): 1 = Non Problematic 5 = Likely to Prevent Access

Diabetes

Summary of definition and information by Diabetes UK (www.diabetes.org.uk):

Diabetes is a common condition which is either classed as Type 1 or Type 2 and results is high levels of blood glucose (sugar) as the body cannot make a hormone called insulin. Type 1 diabetes is a serious and lifelong condition and complications can result in damage to the eyes, heart, kidneys and feet.

Type 2 diabetes is more common amongst suffers and can be lifelong, it is easier to treat than Type 1 diabetes and with appropriate action suffers can live a near normal life.

(Continued)

Diabetes has a high prevalence in the population when compared to other disabilities and particularly with type I diabetes, including the need to treat this with insulin ensures that this is recognised as a disability under the Equality Act 2010.

The research identified diabetes as a condition reported by those with other disabilities and not as a 'stand-alone' condition. Therefore, it is difficult to ascertain disability-specific requirements for accessing commercial property, goods, services and employment.

Down's Syndrome

Summary of the definition and information by The National Health Service [NHS] (www.nhs.uk/conditions/downs-syndrome/):

Down's syndrome is a congenital condition which means it is present at birth and is not due to any actions of the parents before or during the pregnancy. There is a variance or severity in the condition which can result in a range of learning disability, however, some with the condition are able to be more independent while others may need increased regular care or assistance.

People with Down's syndrome are like everyone in the sense that they have their own personalities as well as like and dislikes, this and other characteristics are essentially what makes them who they are.

Children with Down's syndrome have some level of learning disability and need help accordingly, they may need to see a speech and language therapist to help with speaking or seek advice for hearing as well as possible physiotherapist help where they have low muscle tone.

There are other possible health conditions associated with Down's Syndrome and these include:

* increased susceptibility to infections such as pneumonia or flu;
* heart conditions;
* autism and
* developing dementia at a younger age.

There were insufficient responses to the survey data request in the research concerning the disability-specific requirements for those with Down's Syndrome. While there may be specific access requirements associated with this condition, it is important to note that it is necessary to consider the other possible health conditions. Many of these have been analysed specifically in this chapter, and when contextualised with the needs of those with Down's Syndrome, this adds another layer of accessibility requirement.

Dwarfism – Restricted Growth

Summary of definition and information by The National Health Service [NHS] (www.nhs.uk/conditions/restricted-growth/):

The term Restricted Growth is sometimes known as dwarfism and presents as individuals with unusually short height. It is recognised that the two main types of restrictive growth are attributable to proportionate short stature (PSS) or disproportionate short stature (DSS).

As indicated by the term proportionate in PSS, this relates to a general and overall lack of growth, whereas disproportionate relates particularly to the arms and legs but can also present as bowed legs or curvature of the spine.

Those with restricted growth or dwarfism can have in many respects long and normal lives.

There were insufficient responses to the survey data request in the research concerning the disability-specific requirements for those with dwarfism or restricted growth. It is therefore not possible to ascertain any concerns over accessing commercial property, goods, services or employment.

(Continued)

Dyslexia

Summary of the definition and information by the British Dyslexia Association (www.bdadyslexia.org.uk):

Dyslexia is wide-ranging neurological condition affecting reading and writing. It can affect organisational skills, as well as the ability to process information. It is a life-long condition and often is prevalent in family generations.

It is a condition that is widely recognised and supported from primary to higher education. It can be managed with targeted support in line with individual requirements. It should be noted that dyslexic people may have other strengths, and they are often creative individuals.

In accordance with the definition of disability contained within The Equality Act 2010, dyslexia is a long-term condition that has a substantial effect on the ability to undertake normal day-to-day activities. However, this is in the context of a learning environment, and schools are obliged to provide the necessary and reasonable adjustment to teaching and learning materials, as well as methods of assessment.

There is no exact defined information concerning the prevalence of dyslexia, and it is believed to be around 10% of the population.

There were insufficient respondents in the research data to establish the disability-specific needs of those with dyslexia in accessing commercial buildings, goods, services and employment. Noted were a small number of respondents who identified as having dyslexia, but this was substantially lower than the 10% suggested by the British Dyslexia Association. Dyslexia was also noted to be reported by those also recognising as having other disabilities and not as a stand-alone condition.

Dyspraxia

Summary of definition and information by the Dyspraxia Foundation (https://dyspraxiafoundation.org.uk/):

Dyspraxia is a wide-ranging condition that affects movement and coordination. It is a hidden disability that typically presents as awkward movements and poor spatial awareness. People with dyspraxia can have poor organisational skills and express difficulty with motor skills.

It is also known as development coordination disorder (DCD). This can affect speech and language, which can restrict ability to keep up with a conversation or difficulty in coordinating the sounds for clear speech.

People with dyspraxia can also have other conditions associated with specific learning disabilities. As a recognised medical disorder, the condition is believed to affect about 10% of the population with 2–4% seriously affected.

As with many disabilities, there is a scale of severity, and in order to justify the provision for reasonable adjustment, there is a need to assess the long-term nature of the condition, as well as the effect on an individual to carry out normal day-to-day activities. It is anticipated that the significance of dyspraxia is more likely to be relevant in a learning environment such as schools, and accordingly, it is anticipated that appropriate reasonable adjustment is made to teaching, learning materials and methods of assessment.

The survey data contained in the research received insufficient respondents reporting as having dyspraxia, with the actual number lower than the 10% cited by the Dyspraxia Foundation. Where evident amongst respondents, this was additional to other disabilities.

(Continued)

It is not possible to ascertain the disability-specific requirements for those with dyspraxia accessing commercial property, goods, services or employment. However, it is anticipated that physical access requirements are probably not too significant. However, the actual services provision, such as teaching and learning at schools or allocating work tasks to someone with dyspraxia in a business, may require a degree of adjustment.

Dystonia

Summary of definition and information by Dystonia UK (www.dystonia.org.uk):

As a neurological movement disorder, dystonia has symptoms of uncontrollable muscle spasms, which can affect any part of the body including the face, torso and limbs. It is triggered by incorrect signals from the brain and is a relatively common movement disorder.

There are estimated to be 100,000 people affected by dystonia, and it is the third most common movement disorder behind Parkinson's and essential tremor. The condition may affect both adults and children; it can typically relate to one or two parts of the body for adults, but when symptoms start in childhood, this can affect multiple parts.

Presently, there is no cure for dystonia, and the spasms can cause pain. It is considered lifelong, but for many, not life limiting. The condition can be inherited and caused by mutations to a specific gene or acquired as a drug reaction, brain injury or another neurological/metabolic disorder. In some instances, there may be no known or clear cause of dystonia.

There were insufficient responses to the survey request from those recognised with having dystonia. Accordingly, it is not known of any disability-specific access requirements. It is evident from the information available at Dystonia UK that this is a neurological condition that presents as a physical impairment. As with many other physical impairments, and depending on the severity of the condition, the requirements could align with those of others in the category of mobility.

Ehlers-Danlos Syndrome

Summary of the definition and information by Ehlers-Danlos Support UK (www.ehlers-danlos.org):

The Ehlers-Danlos syndrome (EDS) are a collective of individual genetic conditions that affect the connective tissue in the body. With this condition, individuals may be hypermobile, bendy or double-jointed. As a consequence, some of the key symptoms include:

- long-term pain;
- chronic fatigue;
- dizziness;
- palpitations and
- digestive disorders.

EDS are complex syndromes with varying degrees of severity in the symptoms presented by individuals. EDS is considered to be a hidden disability. It is also recognised that there is no cure, but many people learn over time how to control it and live full, active lives.

The fatigue of those with Ehlers-Danlos Syndrome is similar to those with ME/chronic fatigue syndrome, and although there is insufficient data in the research findings to categorically identify trends in any disability-specific access requirements, there is another similarity with ME in that there is a lower-than-average frequency in accessing commercial property, goods and services. This may be as a direct result of chronic fatigue but should also recognise that those with EDS also suffer from chronic pain. This can be managed with medication or reduced by physiotherapy or by using supports such as braces or splints. As alluded to previously, there is noted in the research findings, insufficient data or survey responses to categorically assess the disability-specific needs of those with EDS but

(Continued)

based upon the information provided by Ehlers-Danlos Support UK, this appears mainly to be a disability classified to affect mobility alongside dexterity and some other hidden symptoms.

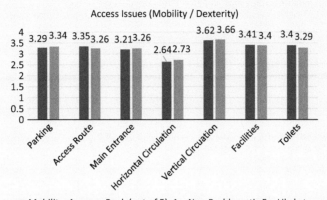

Access Issues (Mobility / Dexterity)

- Mobility: Average Rank (out of 5): 1 = Non Problematic 5 = Likely to Prevent Access
- Dexterity: Average Rank (out of 5): 1 = Non Problematic 5 = Likely to Prevent Access 2

Epilepsy

Summary of the definition and information by the National Health Service [NHS] (www.nhs.uk/conditions/epilepsy/):

Epilepsy is a common condition that affects the brain and causes frequent seizures. These are bursts of electrical activity in the brain that temporarily affect how it works. People with epilepsy can have a variety of symptoms, and the condition is considered lifelong but sometimes can get better with time.

Seizures present as uncontrollable jerking and shaking, which is known as a fit, but the condition can also result in an individual 'blanking out' or becoming stiff. Some sufferers experience strange sensations 'rising' from inside or unusual smells and tastes as well as tingling feelings in the arms or legs. A consequence of epilepsy is passing out and not being aware or remembering the event.

The condition can be treated with medication or medical procedure, as well as with a special diet, but this may vary on individual circumstances. There was noted limited response from the research survey data regarding the disability-specific access requirements of those with epilepsy. While this is no doubt a physical impairment that is long-term and can affect the ability to undertake normal day-to-day activities, the severity and frequency of seizures may influence momentary access to commercial property, goods and services. However, where epilepsy is the sole disability, it is unlikely that this will affect long-term mobility, stamina, breathing or dexterity. Hence, a lack of specific access requirement aligns to the notion that those with managed epilepsy and controlled seizures can lead normal lives.

Fibromyalgia

Summary of the definition and information by the National Health Services [NHS] (www.nhs.uk/conditions/fibromyalgia/):

Fibromyalgia, also called fibromyalgia syndrome (FMS), is a long-term condition that causes pain all over the body as symptoms such as:

- increased sensitivity to pain;
- stiffening of the muscles;

- sleep difficulties (leading to fatigue and tiredness);
- mental processing, headaches;
- irritable bowel syndrome (IBS) and
- feelings of frustration, low mood or worry.

There is a recognition that the exact cause of the condition is unknown, and at present, there is no cure but a number of different treatments. There is also a notion that this could affect 1 in 20 people, but because it is difficult to diagnose, the exact prevalence is unknown.

The findings of the research and identified some but limited response from those reporting fibromyalgia. In all cases, this was not reported as a stand-alone disability but additional to other physical or mental impairments. It is therefore difficult to pinpoint any disability-specific access requirements.

Based upon the symptoms, it can be stated that those who identify as having increased sensitivity to pain or stiffening of muscles may present as requiring access considerations that align within the mobility classification of conditions.

Furthermore, fatigue as a symptom could mean that this is also classed as a stamina condition, and typically, the distance of access routes and vertical circulation may be of concern.

Finally, there is a recognition that those with irritable bowel syndrome (IBS) may consider the provision of sanitary facilities as a high priority when accessing commercial buildings, goods, services and employment.

Friedreich's Ataxia

Summary of the definition and information by the National Health Service [NHS] (www.nhs.uk/conditions/ataxia/symptoms/):

Friedreich's ataxia is the most common type of hereditary ataxia and the symptoms can include:

- problems with balance or coordination;
- speech problem (slurring words, slow or unclear speech);
- weakening in the legs and eventual wheelchair use;
- difficulty in swallowing;
- some abnormal curvature of the spine as seen with scoliosis and
- sight and vision loss as well as diabetes.

The condition can cause cardiac and breathing difficulties, chest pain and loss of sensation to the hands and feet; therefore, this is a complex condition with a multiple of possible symptoms.

In the terms of contributing to physical disability, Friedreich's Ataxia can cover:

- mobility;
- stamina and breathing;
- vision and
- hearing.

These categories relate to those that require the most disability-specific access requirements. However, there were insufficient responses to the survey data request in the research concerning the actual individual disability-specific requirements for those with Friedreich's ataxia.

Huntington's Disease

Summary of the definition and information from the Huntington's Disease Association (www.hda.org.uk):

Huntington's disease is genetic and progressive condition with presently no cure. It affects the nervous system in brain and spinal tissue, which is used for coordination. The symptoms include changes in

(Continued)

movement as well as learning, thinking and emotional behaviour. The disease can occur in all ages, but it is often between the ages of 30 and 50; as it is not presently curable, the symptoms have to be managed to improve quality of life.

There are three defined stages of the disease, with a range of symptoms, but in essence the condition affects:

- Early stages: unexpected movements and experiencing difficulty doing things like turning the pages of a book, task take longer to achieve, experiencing difficulty with planning and thinking, becoming angry/frustrated or irritable.
- Middle stages: involuntary contractions of muscles causing movement and stiffness, slowness of movement and increased clumsiness, speech impairment, difficulty in swallowing and intensifying feelings or frustration or anger;
- Later stages: experiencing weight loss and lack of nutrition, speech and swallowing problems, issues with movement or stiffness and difficulty in communication.

It should be noted that those with Huntingdon's may present with a varying degree of the symptoms listed.

Noted in the research were no reported incidence of Huntingdon's amongst the respondents. While it is not possible to therefore detail any disability-specific access requirements, the symptoms identified align with physical impairments associated within the mobility, stamina and breathing categories as well dexterity.

Inflammatory Bowel Disease (Chron's Disease/Ulcerative Colitis)

Summary of the definition and information provided by the National Health Service [NHS] (www.nhs. uk/conditions/inflammatory-bowel-disease/):

Inflammatory bowel disease (IBD) is a term used to describe conditions that cause severe tummy pain and diarrhoea. IBD is long-term, but there are treatments that can help with the symptoms. The main types of IBD are Crohn's disease and ulcerative colitis. It should be noted that IBD is different from irritable bowel syndrome (IBS), even though some of the symptoms may be similar. In essence, the main symptoms of inflammatory bowel disease (IBD) can include:

- diarrhoea that lasts longer than four weeks;
- tummy pain;
- blood or mucus (clear slime) in poo;
- bleeding from the bottom;
- feeling tired all the time and
- losing weight without trying.

Treatment for inflammatory bowel disease (IBD) will depend on the symptoms, and for the very sick, it may need to be treated in hospital; accordingly, the treatments may include:

- medicines to help ease symptoms, such as steroids, and other medicines that can help keep symptoms under control and
- surgery to remove part of the bowel, if symptoms are severe.

Noted was a limited response within the data collection for the research regarding those with IBD and disability-specific access requirements. With the limited data set, it was noted that those with this condition consider it to be a hidden disability, and the following access requirements have been noted:

(Continued)

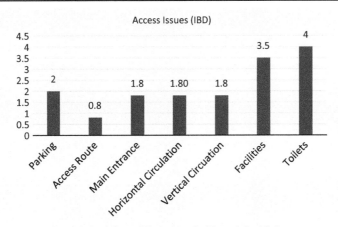

Access Issues (IBD)

■ Average Rank (out of 5): 1 = Non Problematic 5 = Likely to Prevent Access

Kyphosis (Scheuermann's kyphosis)

Summary of the definition and information from the National Health Service [NHS] (www.nhs.uk/conditions/kyphosis/):

Kyphosis is curvature of the spine that causes the top of the back to appear more rounded than normal. The symptoms may be only visual in the sense that there is a hunched appearance, but there are instances where the condition has other symptoms such as:

• back pain and stiffness;
• tenderness of the spine and
• tiredness.

As the condition worsens over time, this may lead to difficulties in breathing and eating. This occurs when there is some excessive curvature of the central section of the vertebral column.

This can be as a result of activity such as posture or continued carrying of heavy items or even as the result of a spinal injury. It can also be caused by abnormal development in the womb during early development stages and also curvature of the spine can increase with old age towards the latter years of life.

This is no doubt a physical impairment which is long-term in nature and affects an individual's ability to carry out normal day-to-day activities. Noted was insufficient data in the responses contained in the research to develop an understanding of disability-specific access requirements. However, based upon the definition of the condition, it is likely that those with Kyphosis will experience similar access concerns to those with mobility, stamina and breathing classifications of disability.

Learning Disability

Summary of the definition and information by the National Health Service [NHS] – www.nhs.uk/conditions/learning-disabilities/):

A learning disability is an umbrella term that can be used to describe or include specific learning disabilities, such as dyslexia or dyspraxia and the essentially affects the way in which a person may learn things throughout their life. Both dyslexia and dyspraxia have been covered independently as separated disabilities.

There is some debate on whether learning disabilities are actually a disability, and it is possible that some with specific conditions contained within this general term would not consider themselves to be disabled. However, as with all disabilities, it is necessary to return to the default analysis on whether this is a physical or mental impairment that is long term and has a substantial effect on an individual's ability to carry out normal day-to-day activities.

(Continued)

The severity of learning disabilities can be varied and are often unique to individuals. In essence, this may present as difficulty in understanding complicated information or learning some skills. A person with a learning disability may have difficulty in looking after themselves or being able to live alone.

Many individuals with a learning disability can be employed or work, have relationships, live alone, and obtain qualifications, while others may need continued support.

In the context of the research undertaken into disability-specific access needs and learning disabilities, it is important to note that the research focused on the physical access to commercial buildings, goods, services and employment. It is not concerned with the obligations of the products delivered by service providers. Therefore, disability-specific access requirements in this instance do not seek to analyse specific learning material or teaching support in the context of education. The survey data received negligible responses from those recognising as having a learning disability. While the significance of this group of conditions should not be underestimated in the terms of intellectual or personal development, it is unlikely that they present access difficulties. An exception to this may be where a learning disability is additional to another disability that does present with access problems.

Lupus

Summary of the definition and information at Lupus UK (www.lupusuk.org.uk):

Lupus is a chronic autoimmune disease in which the body's own immune system attacks its own tissues causing inflammation and damage. In some cases, this affects organs in the body and joints. The majority of those with lupus suffer from fatigue which can also have a substantial effect of carrying out daily activities. It is recognised that lupus is a complex condition in which individuals can have one or more additional conditions.

It is considered that two of the more common symptoms include:

- pain to joints/muscles and
- fatigue.

Despite the research recording insufficient responses from those recognising with the condition, it is evident from the two key symptoms and also similar symptoms identified be different disabilities that this is within the categories of disability affecting mobility, stamina and breathing. Accordingly, the provision of accessible parking, access routes and vertical circulation to and from buildings is likely to be of significance.

According to Lupus UK, there are other potential symptoms, and although wide ranging, it is important to note that not all those with lupus have exactly the same symptoms which include:

- rashes;
- anaemia;
- light-sensitivity;
- headaches/migraines;
- hair loss;
- oral/nasal ulcers;
- brain fog;
- depression and
- anxiety.

Marfan Syndrome

Summary of the definition and information from the Marfan Trust (www.marfantrust.org) and the National Health Service {NHS] (www.nhs.uk/conditions/marfan-syndrome/):

Marfan Syndrome is an inherited disorder affecting the connective tissue in the body that leads to medical problems affecting the heart, eyes and skeleton. It is recognised that Marfan Syndrome and related conditions can shorten lives, especially if left untreated.

These conditions can have a significant impact on quality of life and the ability to carry out normal day-to-day activities; however, with regular medical intervention and advances, it is also recognised that individuals with Marfan Syndrome can life long and productive lives.

The symptoms of Marfan Syndrome are varied and fall within the following categories:

* skeleton;
* eyes and
* cardio vascular.

Problems affecting the skeleton and bones include curvature of the spine, backache, headaches and numbness or pain in the legs. Eye conditions include short sightedness, glaucoma, cataracts and retinal detachment; additionally, there can be some quite complex heat conditions associated with Marfan Syndrome.

This is a notion that this is quite a rare condition, and the research survey data did not contain sufficient data to identify any trends associated with disability-specific requirements for the access of commercial property, goods, services and employment. However, the very nature of the potential symptoms affecting the skeleton, eyes and heart means that this is likely to place the disability into the following categories:

* mobility;
* stamina/breathing and
* vision.

These three classifications of disability are more likely to generate significant requirements for accessing commercial property, goods, services and employment.

Mental Health

Summary of the definition and information by mind is an umbrella term to describe a wide range of conditions that can include anxiety or depression through to more complex diagnoses, which may include bipolar disorder or schizophrenia. There is a variance between the conditions, and the symptoms may be different between individuals.

Mental health is more prevalent amongst those of working age but has been on the increase with children. In the terms of a recognised mental impairment under the Equality Act 2010 and government statistics concerning disability, there has been an increase in those recognising as suffering from adverse mental health issues.

There are many effective treatments for mental health, which means those diagnosed with the varying condition can lead productive lives.

The research addressing disability-specific access respondent noted a number of respondents who confirmed to be suffering from a mental health issue. In line with the information available via Mind, this was noted to be prevalent amongst working-age respondents. It was also noted that this was a stand-alone disability, as all of the respondents identified mental health additional to other reported disabilities. Accordingly, it is difficult to establish any disability-specific issues regarding access to commercial property, goods, services and employment.

There is no doubt that the symptoms associated with mental health can be debilitating, and this can be particularly challenging in a work environment, but it is necessary to note that the research concerned the physical experiences of those accessing commercial buildings, goods, services and employment. It is not concerned with the allowance in working practices of businesses and the reasonable adjustments made for those with mental health conditions.

(Continued)

It may be suggested that many mental health conditions are hidden disabilities, and where this is the sole disability of an individual, it is unlikely that this will result in the same issues or access needs of those who categorise their disabilities affecting mobility, stamina and breathing or vision and hearing.

Motor Neurone Disease (MND)

Summary of the definition and information from the Motor Neurone Disease Association (www. mndassociation.org):

MND is the abbreviation for motor neurone disease, which affects the nerves in the brain and spine. The function of the nerves is to control the movement of muscles to many parts of the body. There are essentially four recognised types of MND, and it is a progressive condition with the symptoms worsening with the duration of time.

Accordingly, there are a wide range of symptoms, and it is recognised that the overlap in symptoms can introduce difficulty in differentiating between the types of disease.

Those with MND typically have symptoms resulting in the weakening, stiffening or wastage of muscles. This can affect the ability to:

- walk;
- talk;
- eat and
- breathe.

It is a debilitating disability with poor outcome, and presently, there is no known cure; it is life-shortening. Medication can be used to treat the symptoms to manage and enhance quality of life.

There was insufficient research data received concerning the individual disability-specific access requirements for those with MND. Despite this, there is a notion that the symptoms suggest that this disability primarily sits within the following categories:

- mobility,
- stamina and breathing and
- dexterity.

In the context of access to commercial property, goods, services or employment, it therefore appears that those with MND will experience significant challenges with most aspect of access from the provision of accessible parking through to the access routes, entry of buildings and horizontal circulation. The provision of facilities and in particular sanitary provision is important for those with this condition.

Multiple Sclerosis (MS)

Summary of the definition and information from the Multiple Sclerosis Society (www.mssociety.org.uk):

Multiple sclerosis (MS) is a condition that affects the brain and spinal cord, it has a variety of different symptoms which include:

- fatigue;
- numbness and tingling;
- loss of balance and dizziness;
- stiffness or spasms;
- tremor;
- pain;
- bladder and bowel problems;
- vision problems and
- problems with memory and thinking.

(Continued)

It is also recognised that the severity of symptoms can vary as well as the fact that these can come and go with time.

Presently there is no known cure for MS, but several treatments can be used to manage the condition.

Minimal responses were noted in the research data from those recognising with this condition, and the response rate was too low to verify specific disability requirements in accessing goods, services and employment. However, it was noted that all respondents use mobility aids such as wheelchairs, walking sticks/crutches or rollators. All respondents also classify their disability to affect mobility, stamina & breathing as well as dexterity. There is however enough research response to identify the trend in general access requirements per classification:

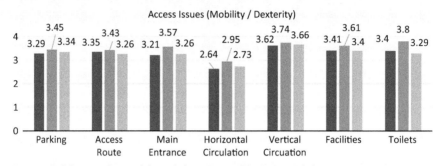

Access Issues (Mobility / Dexterity)

- Mobility: Average Rank (out of 5): 1 = Non Problematic 5 = Likely to Prevent Access
- Stamina & Breathing: Average Rank (out of 5): 1 = Non Problematic 5 = Likely to Prevent Access
- Dexterity: Average Rank (out of 5): 1 = Non Problematic 5 = Likely to Prevent Access

Multiple System Atrophy

Summary of the definition and information by the National Health Service [NHS] (www.nhs.uk/conditions/multiple-system-atrophy/):

Multiple system atrophy (MSA) is a rare condition of the nervous system that causes gradual damage to nerve cells in the brain. It has a wide variety of symptoms, but not all individuals present with all of these. The condition affects movement, balance and the automatic nervous system. It has been observed that the symptoms are similar to Parkinson's disease, and this can also affect the basic functions of breathing, bladder control and digestion.

In line with the rare occurrence of this condition, there were no responses in the research data to those with MSA, and when attempting to assess the disability-specific access requirements of the condition, It is necessary to recognise the following symptoms:

- bladder problems;
- low blood pressure;
- reduced speed of movement and stiffness;
- shoulder and neck pain;
- muscle weakness and
- blurred vision.

All of these symptoms will no doubt affect the ability to access commercial property, goods, services and employment. It needs to be stated that as with many disabilities, the number, combination,and severity of symptoms have an influence on access requirements. Based upon the listed, relevant symptoms these appear to align this disability within the following categories:

- mobility;
- breathing and stamina;

(Continued)

- dexterity and
- vision.

As with disabilities grouped into these classifications, it is evident that there will be significant access concerns.

Muscular Dystrophy

Summary of definition and information from the National Health Service [NHS] (www.nhs.uk/conditions/muscular-dystrophy/):

Muscular dystrophy is a term used to describe a collective of inherited generic conditions that present as muscle weakness. Muscular dystrophies (MD) are progressive conditions that result in increased levels of physical disability. Presently, there is no cure for MD, although treatment can help to manage the symptoms.

There are a number of different types of muscular dystrophy, and some of these can affect the heart or even the group of muscles which control breathing, in these instances, the condition can be life threatening.

Treatment for muscular dystrophy comprises mobility assistance such as exercises, physiotherapy or mobility aids. According to the NHS, treatment can include corrective surgery or prescribed medicine to treat the symptoms.

The results from the research indicated insufficient returns from those recognising as being diagnosed with muscular dystrophy. Accordingly, it has been noted that the symptoms align with disabilities that are classified as:

- mobility;
- stamina and breathing and
- dexterity.

This general-access specific requirements associated with this classification of disability score highly regarding the lack of access provision likely to prevent access to commercial buildings, goods, services and employment.

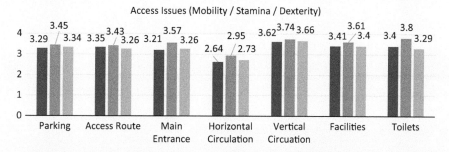

Access Issues (Mobility / Stamina / Dexterity)

■ Mobility: Average Rank (out of 5): 1 = Non Problematic 5 = Likely to Prevent Access
■ Stamina & Breathing: Average Rank (out of 5): 1 = Non Problematic 5 = Likely to Prevent Access
■ Dexterity: Average Rank (out of 5): 1 = Non Problematic 5 = Likely to Prevent Access

Osteoporosis

Summary of the definition and information provided by the National Health Services [NHS] (www.nhs.uk/conditions/osteoporosis/):

(Continued)

Osteoporosis is a condition which results in the weakening of bones in the body making them more susceptible to breaking. It is a condition which develops over time and is typically (but not always) associated with old age.

Typically, common injuries associated with osteoporosis are:

- broken wrists;
- broken hips or hip fractures and
- broken vertebrae in the spine.

However, broken bones may occur to other parts of the body.

The research findings identified a number of respondents with osteoporosis, and when analysing the data, it was noted a greater ratio of female to males with the condition. Over 50% of those with the condition confirmed that they were 65 years old or over.

There was also noted some overlap with other reported disabilities, including mental health and arthritis. The general trend is disability access requirement for those with osteoporosis aligns to an extent with mobility issues.

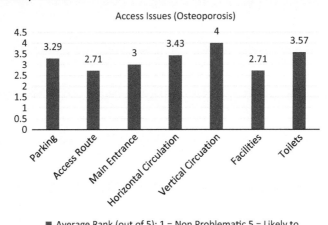

Access Issues (Osteoporosis)

■ Average Rank (out of 5): 1 = Non Problematic 5 = Likely to Prevent Access

Parkinson's Disease

Summary of the definition and information by Parkinson's UK (www.parkinsons.org.uk):

Parkinson's is a progressive neurological condition which worsens with time and is recognised as one of the fastest growing neurological conditions worldwide.

The main symptoms of the condition are a tremor or shaking and also reduced speed of movement characterised by a shuffled walk. Parkinson's is caused when the nerve cells that make dopamine die, and this affects messages sent from the brain that coordinate movement. The symptoms of Parkinson's are:

- tremor;
- rigidity or stiffness;
- reduced speed in movement;
- memory and thinking problems;
- sleep disorder;
- pain and
- mental health such as anxiety and depression.

(Continued)

It is recognised that those with Parkinson's do not always present all of the symptoms, and these are categorised as those affecting movement or balance (known as motor symptoms) and unseen symptoms (non-motor) associated with pain, sleep issues and mental health).

Despite there being a high prevalence within the population, there was noted insufficient responses within the research to establish trends in disability-specific access requirements for accessing commercial property, goods, services and employment.

However, through analysis of the information detailed concerning the symptoms of the condition, this appears to suggest that this disability is located within the classifications of mobility and dexterity. It has been established that these two categories have significant access requirements concerning all aspects of accessing commercial buildings, goods, services and employment.

Polio

Summary of the definition and information from the National Health Service [NHS] (www.nhs.uk/conditions/polio/):

Polio is a serious infection that is now considered to be very rare in the UK with the last reported case in 1984 and it has been eradicated from Europe since 2003. This is due to an effective vaccination programme, although there are some parts of the World where polio is still present.

It is a virus that is noted to have the following symptoms:

- high temperature;
- extreme tiredness and fatigue;
- headaches;
- vomiting;
- neck stiffness and
- muscle pain.

While the symptoms last for around 10 days, more serious and rare symptoms can affect the brain as well as the nerves causing muscle weakness and paralysis in the legs. It is noted that most people recover from polio but for some, they can be left with permanent disability.

In line with the rare nature of this medical condition and any obvious absence of those who have contracted this in the UK, there was noted no responses in the research data to establish any disability specific access requirements.

Scleroderma

Summary of the definition and information by the National Health Service [NHS] (www.nhs.uk/conditions/scleroderma/):

Scleroderma is a term which covers a variety of conditions that affect the immune system and accordingly referred to as autoimmune conditions. The symptoms of scleroderma primarily affect the skin and can cause the hardening or thickening. It can sometimes cause problems with muscles, bones, internal organs and blood vessels.

There is an understanding that the condition is associated with an overactive immune system, which appears to be out of control, and this causes thickening and scarring of tissue, known as localised scleroderma.

There can be more significant symptoms associated with systematic scleroderma, which can affect internal organs and lead to severe or life-threatening problems. Some of the other symptoms of diffuse systematic sclerosis are:

- weight loss;
- fatigue and
- joint pain or stiffness.

Scleroderma can be treated with both medications to address circulation and reduce the activity of the immune system to attenuate the progression of the condition. Medication can also be used to address joint and muscle complaints with a number of treatments that can be applied to the surface of skin. It is necessary to monitor the condition with regular testing to establish any problems with organs.

The research findings noted insufficient responses in the data received concerning the disability-specific requirements for accessing commercial buildings, goods, services and employment. Based upon the analysis of the symptoms, access issues may affect mobility, stamina and breathing, but this is likely to be dependent upon the severity of the conditions. It is noted that mobility, stamina and breathing issues have significant consequences on most of the recognised access provisions.

Scoliosis

Summary of the definition and information from the Scoliosis Association UK (www.sauk.org.uk):

Scoliosis is a physical disability that results in a curvature or twisting of the spine. It is not contagious or related to something someone has or has not done. Scoliosis can occur at all stages of life and is sometimes related to neuromuscular conditions such as muscular dystrophy or cerebral palsy.

Scoliosis can also develop as part of a syndrome, such as Marfan Syndrome. The twisting or curvature of the spine can also cause misalign connecting ribs as well as possible forcing a shoulder blade to stick out or create an uneven waist.

The condition can occur before birth and at all stages in life from childhood (early onset), through the teenage years and into adulthood. It is a physical impairment and according to the data contained in the research findings, it was noted that the condition was more prevalent from middle age and classified by those with it to affect mobility and dexterity.

Accordingly, all aspects of accessing commercial buildings, goods, services and employment are considered to be problematic with the exception being horizontal circulation, which is considered to be a lesser issue by those recognising as having scoliosis.

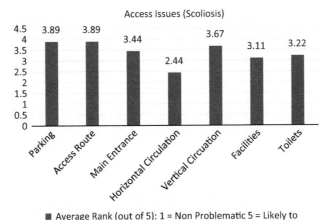

Access Issues (Scoliosis)

■ Average Rank (out of 5): 1 = Non Problematic 5 = Likely to Prevent Access

(Continued)

Sjögren's Syndrome

Summary of definition and information by the Sjögren's Foundation (www.sjogrens.org):

Sjögren's is a systemic autoimmune disease that affects the entire body and the symptoms may vary and get increasing worse or stay the same, these include:

- extensive dryness;
- chronic fatigue;
- pain;
- major organ involvement;
- neuropathies and
- lymphomas.

There are degrees of severity experienced by individuals with the condition, and while this may be mild for some, it can be life-limiting for others. About half the cases of Sjögren's occur with other autoimmune connective-tissue disease, such as Rheumatoid Arthritis, Lupus or Scleroderma.

The research findings noted some but insufficient responses to the research data collection concerning Sjögren's Syndrome, and it was noted that this was combined with other disabilities. There is no doubt that some of the more severe physical symptoms contribute to classifying this as a disability affecting primarily mobility, but also the notion of chronic fatigue makes those with this condition likely to consider this to affect breathing and stamina too.

Other disabilities identifying within the classifications of mobility and in particular stamina note significant difficulties when accessing commercial properties, goods, services and employment.

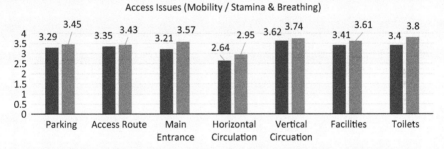

Access Issues (Mobility / Stamina & Breathing)

■ Mobility: Average Rank (out of 5): 1 = Non Problematic 5 = Likely to Prevent Access

■ Stamina & Breathing: Average Rank (out of 5): 1 = Non Problematic 5 = Likely to Prevent Access

Spina bifida

Summary of the definition and information by the National Health Service [NHS] (www.nhs.uk/conditions/spina-bifida/):

Spina bifida occurs during the development of a baby in the womb when the spinal cord does not form properly, leaving a gap. When the neural tube does not develop or close properly, this can lead to issues with the spinal cord and the associated vertebrae.

There are number of different variants of spina bifida, which relate to differences in the severity of the symptoms. It should be noted that in most cases, corrective surgery can be performed to close the spinal opening, but where there has been damage to the nervous system, this can result in the following symptoms:

(Continued)

- complete paralysis or weakness of the legs;
- bowel and urinary incontinence and
- fluid on the brain for babies.

It should be noted that most people with the condition have normal intelligence but some have learning difficulties.

Noted in the research data findings were no responses from those identifying as having spina bifida. However, with weakness or paralysis of the legs, the treatment to facilitate mobility includes wheelchairs, walking supports and braces or walking sticks and crutches.

There is sufficient data within the research findings to identify trends in the access concerns of those who are either powered or manual wheelchair users or are ambulant disabled utilising walking supports, crutches and walking sticks.

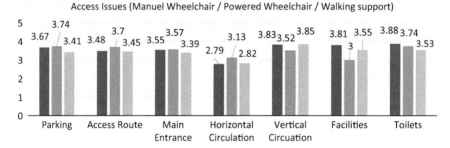

Access Issues (Manuel Wheelchair / Powered Wheelchair / Walking support)

- Manual Wheelchair: Average Rank (out of 5): 1 = Non Problematic 5 = Likely to Prevent Access
- Powered Wheelchair: Average Rank (out of 5): 1 = Non Problematic 5 = Likely to Prevent Access
- Walking support: Average Rank (out of 5): 1 = Non Problematic 5 = Likely to Prevent Access

Spinal Injury

Summary of definition and information provided by the Spinal Injuries Association (www.spinal.co.uk):

A spinal cord injury can occur through an accident, illness or health condition and the consequence of this can have profound long lasting effects.

The spinal cord is effectively the link between the brain and the rest of the body. It communicates instructions on movement and is a two-way by virtue that is also paramount in communicating physical feelings or sensations.

Damage to the spinal cord will result in permanent physical impairment with the consequences increasing if this occurs in the neck region. A spinal cord injury to the back results in paraplegia, affecting movement or sensation in the legs and abdominal muscles. Damage to the spinal cord in the neck results in tetraplegia, which affects all limbs as well as abdomen and chest muscles.

The result of a spinal injury can be profound, and while the research results received insufficient data from those with this condition in relationship to accessing commercial property goods, services and employment, this no doubt is classified as primarily a disability affecting movement. It should be concluded that as a disability that has an effect on the ability to carry out normal day-to-day activities, this can also alter bowel and bladder function. Those with a spinal cord injury are likely to be

(Continued)

confined to a wheelchair, and accordingly, this results in a high level of concern regarding access. What is evident in the granular detail of the research and illustrated in Chapter 9 are relatively simple reasonable adjustments such as the placement of a dropped kerb adjacent to accessible parking spaces or on access routes. The absence of these can make a profound negative impact upon accessibility.

Spinal Muscular Atrophy

Summary of the definition and information provided by Muscular Dystrophy UK (www. musculardystrophyuk.org/conditions/spinal-muscular-atrophy):

Spinal Muscular Atrophy (SMA) is a rare, genetic neuromuscular condition causing progressive muscle wasting (atrophy) and weakness leading to loss of movement. This may affect:

- crawling and walking ability;
- arm, hand, head and neck movement and
- breathing and swallowing.

The conditions and disability-specific access requirements have been considered more generally for muscular dystrophy previously in this chapter and in line with this SMA presents a disability that is within the mobility and dexterity classification. Muscle weakness is also likely to affect stamina and contribute to fatigue.

There was noted insufficient data contained within the research identifying disability-specific access requirements of this condition. However, on the basis of the effect on mobility, stamina and dexterity it is evident that every facet of access within the built environment is likely to be problematic.

Stroke

Summary of the definition and information provided by the National Health Service [NHS] (www.nhs. uk/conditions/stroke/) and the Stroke Association (www.stroke.org.uk/effects-of-stroke):

A stroke occurs when the blood supply to part of the brain is cut off, and it is a serious medical condition requiring immediate intervention. The blood supply to the brain is either cut off as a result of blood clot, which is known as an ischaemic stroke or when a blood vessel supplying the brain bursts and this is known as a haemorrhagic stroke.

The consequence of a stroke can be varied with some individuals having physical effects such as:

- muscle weakness affecting one side of the body;
- difficulty in walking;
- muscle tension;
- pain and
- extreme fatigue.

A stroke can cause problems with communication and understanding as well as difficulties in reading and writing. However, from an accessibility perspective, those visiting commercial properties for goods, services or employment mobility will contribute significantly to access requirements. It is estimated that nearly three quarters of those who have a stroke have weakness to their arms and legs. This can have a profound effect on walking or holding things which also affects individual dexterity.

Noted is a variance in the possible symptoms associated with stroke, and there is also a variance in the severity of the symptoms which can result in different personal access requirements. There were a number of responses in the data collection for the research into disability-specific access requirements, but the overall numbers were too low to establish specific trends.

(Continued)

Based upon the symptoms and noting that nearly 75% of those who have had a stroke experience weakening of leg muscles, it is the availability of positioning of accessible car parking spaces and access routes to buildings that are likely to be problematic, this alongside accessing main entrances and vertical circulation around a building. Accessible facilities and in particular sanitary provision are critical which concludes that those with significant disability attributable to the condition have high disability access requirement.

General Disability Classification Access Trends

The experiential understanding of those reporting a physical or mental impairment and how their disability interacts with the built environment has identified some key overall themes.
  The following are some of the key headline findings:

* response rate 67% female and 33% male.
* 45% of respondents over the age of 65.
* 97% of respondents consider themselves to have a long term physical or mental impairment affecting normal day-to-day activity.
* 48% of respondents report to having more than one disability.
* 68% of respondents disagree that commercial properties for accessing goods, services and employment are accessible.

With there being limited insufficient data available for every one of the listed disabilities regarding their specific access concerns likely prevent access to commercial properties, goods and services, the combined data has been placed into the categories of disability:

Category	Parking	Access Route	Main Entrance	Horizontal Circulation	Vertical Circulation	Facilities	Toilets
Mobility (wheelchair)	3.62	3.53	3.51	2.91	3.77	3.60	3.88
Mobility (ambulant)	3.41	3.45	3.39	2.82	3.85	3.55	3.53
Stamina/ Dexterity	3.45	3.43	3.57	2.95	3.74	3.61	3.80
Mental Health	2.95	3.2	3.43	3.24	3.10	3.34	3.57
Memory	3.08	3.46	3.38	2.92	3.54	3.23	3.85
Hearing	3.36	3.45	2.95	2.95	3.59	3.45	3.91
Vision	3.00	3.56	3.53	3.50	3.76	3.44	3.93
Learning				**Insufficient Data**			
Social/Behavioural	3.20	3.20	2.90	3.10	3.40	3.80	3.40
Other (incl hidden)	3.13	3.04	3.35	3.09	3.61	3.54	3.50

In conclusion, there exists a wide variety of different physical and mental impairments than have their own disability-specific requirements for accessing commercial buildings for goods, services and employment. When placed into the context of the different recognised categories of disability, it is evident that all of these appear to have similar access concerns; however, when delving deeper into the research findings, it is noted that close to 50% of respondents have more than one disability, and for example, those with mental health may not be associated with access issues. But those with mental health who also have a physical impairment recognise that the combination of disabilities can have a profound effect on access.

The underlying theme emerging from the analysis of the different disabilities is the notion that designing a 'one size' guidance or framework on providing access to commercial buildings, goods, services and employment is more likely to cover all eventualities rather than adopting a nuanced approach to providing access for specific disabilities.

Notes

1 Fleck, J. (2019). Are you an inclusive designer?. RIBA Publishing.
2 The social model of disability – Sense
3 Family Resources Survey: Financial Year 2020 to 2021 – GOV.UK (www.gov.uk)
4 Covid: Disabled people account for six in 10 deaths in England last year – ONS – BBC News
5 Family Resources Survey: financial year 2020 to 2021 – GOV.UK (www.gov.uk)
6 Hancock, R., Morciano, M., and Pudney, S. (2016). Disability and Poverty in Later Life. Joseph Rowntree Foundation.

References

BBC. (2021). *Covid: Disabled people account for six in 10 deaths in England last year – ONS.* [online] Available at: https://www.bbc.co.uk/news/uk-56033813 [Accessed 14 January 2024].

Fleck, J. (2019). *Are you an inclusive designer?*. RIBA Publishing.

Gov.UK. (2023). *National statistics – Family resources survey: Financial year 2020–2021.* [online] Available at: https://www.gov.uk/government/statistics/family-resources-survey-financial-year-2020-to-2021 [Accessed 14 January. 2024].

Sense. (n.d). *The social model of disability.* [online] Available at: https://www.sense.org.uk/about-us/the-social-model-of-disability/ [Accessed 14 January. 2024].

Chapter 4

Legislation and Standards

Introduction

Concerning the legal framework or standards applicable to creating an inclusive environment to commercial property, it is important to contextualise this book as examining the physical access requirements to premises. It is acknowledged that those providing goods, services or employment are obliged to comply with a range of statutory requirements. Some of these are covered in this chapter, but to fully understand inclusive environments and access to commercial property, goods, services and employment, it is paramount to discuss the barriers and drivers to accessibility. Barriers historically have been presented by the individual disabilities described in Chapter 3, but more so by the societal barriers alluded to in the social model of disability. There is a notion that it is the inability of society to permit physical access to goods, services and employment that create the barriers. Therefore, more emphasis and guidance are required to society or more so employers and those delivering goods as well as services to remove these barriers. However, with a need for a uniform approach in addressing inclusivity, there will always be the requirement for legal definition. Drivers are perhaps more seen as the positive or value that inclusivity can bring to society; too often there is a cost/benefit element associated with doing things in a commercial environment. Therefore, commercial compromise often seeks to do nothing or the relevant legal minimum concerning accessibility. In simple terms, adopting the term 'carrot and stick' to the barriers associated with implementing accessibility means the driver is the carrot or opportunity and stick is the legal requirement or obligation.

As discussed in Chapter 2, early legislation concerning the provision of disabled rights was in the form of the Chronically Sick and Disabled Persons Act 1970 (CSDPA70), which is still in force and celebrated 50 years of existence in 2020.[1] The CSDPA70 has been the subject of a number of amendments but in essence is largely the same piece of legislation and has stood the test of time. This Act was introduced by MP Alf Morris who had lived with his disabled father, a WW1 veteran.

Creating a flow between the history of disability, definitions and the legal framework gives way to the discussion in Chapter 7, which looks at public and commercial attitudes to disability in the built environment. This also introduces the notion of 'experiential understanding'. There is legislation protecting the rights of those with a physical or mental impairment. However, only those with a disability or caring for someone disabled really understand what it is like to access commercial property, goods, services and employment. The rest of society has to rely on the presence of legislation, which at best is

DOI: 10.1201/9781003239901-4

difficult to intellectually navigate, or they need to revert to guidance published by government departments, professional institutions and charities on various aspects of disability.

Therefore, guidance needs to be clear and as unambiguous as possible, particularly considering many employers may not recognise as having a physical or mental impairment. Furthermore, many professionals designing buildings or auditing the existing built environment for building owners, investors and service providers do not recognise as being disabled. The absence of those with a disability either as auditors, designers or employers means they cannot explicitly relate with what it means to be disabled. Therefore, guidance or audits should take into account the experiential understanding of those with a physical or mental impairment. Presently, there are a number of relevant laws as well as some key guidance, concerning the access of commercial properties, goods, services or employment the principal disability legislation including:

- The Chronically Sick and Disabled Persons Act 1970.
- The Health and Safety at Work Act 1974.
- The Employment Rights Act 1996.
- The Equality Act 2010.
- Equality Act 2010 Statutory Code of Practice Employment.
- Building Regulations – Approved Document Part M.
- British Standard 8300.

The Chronically Sick and Disabled Persons Act 1970

The CSDPA70[2] was a bespoke piece of legislation directed at the welfare of those with disability and to an extent built upon the National Assistance Act 1948 which already made some welfare provision for the disabled.[3] The 'new' act in 1970 concerned the following principal areas:

- Welfare and housing.
- Premises open to the public.
- University and school buildings.
- Advisory committees, etc.
- Miscellaneous provisions.

In essence and concerning access to commercial property, goods, services and employment for those with a physical or mental impairment, it is the sections covering public premises and university or school building that are most relevant.

Premises open to the Public.

Sections 4 to 7 of the CSDPA70 cover premises open to the public which are separated into the following distinct areas:

4. Access to, and facilities open to the public.
5. Provision of public sanitary conveniences.
6. Provision of sanitary conveniences at certain premises open to the public.
7. Signs at buildings, etc.

Essentially, Section 4 of the Act states that premises open to the public should make reasonable provision through parking, access and sanitary facilities for those with disabilities. The terms 'reasonable' and 'appropriate provision' are contained within the text which is open to interpretation. There are a number of exclusions in the premises obliged to provide access and facilities based upon the use class of the buildings, with the 'veto' of not obliging any provision if this is not practicable or reasonable.

Subsection 4.1 A of the Act make reference to appropriate provision being *the Code of Practice for Access for the Disabled to Buildings,* accordingly the British Standards Institution code of practice referred to as BS 5810: 1979. British Standards are not a legal document but provide guidance; irrespective of this British Standard BS 5810:1979 was withdrawn in 2001 and replaced with British Standard 8300:2001. It is currently widely accepted that British Standard BS 8300:2018 is current best practice or 'gold standard' in respect of inclusivity. This will be discussed in more significant detail later in this chapter.

Section 5 of the CSDPA70 covers the provision of public (local authority) conveniences and states that these should meet the needs of disabled persons where reasonable and practicable. Subsections in Section 5 indicate the requirement for the placement of appropriate signage, but there is no reference to published guidance on the specification of the facilities. The provision of sanitary conveniences at certain premises open to the public is covered in Section 6 of the act, but again, this is not widespread to all commercial buildings, the requirements also introduce the notion of reasonableness and practicability.

The provision of signage is identified in Section 7 of the act, and in essence, this details that this relates to building access, facilities and sanitary convenience as well as signage for car parking. The detailed requirements of the signage are not specifically identified, and there is no explicit reference to Building Regulations or British Standards.

The CSDPA70 explicitly covers access to university buildings and schools; this is detailed in Section 8 and obliges all schools as well as further and higher education buildings. The act does not default to reasonable or practicable steps and explicitly identifies parking, access to and within buildings, as well as sanitary conveniences. In contrast, Section 8 A applies a similar requirement for offices and other premises, but this includes reverting to reasonable as well as practical implementation relative to the use class.

In summary, the Chronically Sick and Disabled Persons Act 1970 alludes to access provision to commercial properties but not specifically access to goods and services.

The Health and Safety at Work Act 1974

The Health and Safety at Work Act 1974 (HSWA74) is not primarily concerned with the physical access of commercial buildings, goods and services. However, there is provision within this legislation that affects employment. There is an obligation within the HSWA74 for employers to ensure the safety and welfare of their employees as well as visitors to the place of work. While the Act does not specifically refer to disability, there could be people within the workforce who are recognised as having a physical or mental impairment; accordingly, their welfare should be addressed and covered by the legislation. In the context of employee welfare, those with disability may have specific employment needs, and this is covered more by the Equality Act 2010.

The Employment Rights Act 1996

The Employment Rights Act 1996 is not concerned with the physical access requirements to the place of work for those with a disability. It concerns the rights of employees. With respect to requesting flexible working conditions, those with a disability are treated the same as any employee in the sense that they can make a statutory request for flexible working. While the parents of children have been able to request flexible working, this also includes any member of the workforce providing they have worked the required minimum time for an employer. The most appropriate and comprehensive legislation concerning both the access of goods and services, as well as employment for those with a physical and mental impairment is the Equality Act 2010.

It has been established in Chapter 2 that attitudes towards disability changed significantly after WW1 and even more so after WW2. This is epitomised by early legislation directed to support those with physical or mental impairment. Furthermore, there is documented evidence concerning centuries-old recognition of disability with aspects of society reaching out to offer support. However, it was not until 1995 that there was the first bespoke legislation enacted to prevent disability discrimination. The came in the form of the Disability Discrimination Act 1995 (DDA95), which is the forerunner of the current Equality Act 2010. Before analysing the current requirements for access provision under the Equality Act 2010, it is necessary to look back at the forerunner.

The Disability Discrimination Act 1995

The DDA95 was the first piece of legislation directed at those with disabilities. It obliged those providing access to goods and services not to discriminate access provision to those with a physical or mental impairment. The DDA 1995 was an 88-page document, which was organised into the following key sections:

1 Disability (including definition);
2 Employment;
3 Discrimination in other areas (including access to goods and services);
4 Education;
5 Public transport;
6 The National Disability Council;
7 Supplemental and
8 Miscellaneous.

The Act was considered a highly progressive piece of legislation in the mid-1990s, and the fundamental definition of disability was used to shape that of the definition in the subsequent Equality Act 2010. It also set the foundations for the requirements concerning employment as well as access to goods and services.

However, in 2020, the 25-year anniversary of the legislation had a muted response. While there is an overwhelming recognition that the DDA was a 'milestone' moment, those with disability are critical that there are few tangible improvements to their lives.[4]

The Act was brought about in the context of militant action by those with disability demanding equal rights and more than a quarter of a century after the event, there is a simmering discontent with talk of civil disobedience. Part of the narrative is that disability was removed from the forefront of society by austerity and cuts to public spending, which in part coincided with the repealing of the DDA95 and its subsequent replacement with the Equality Act 2010. At a time when the UK was gearing up to host the most successful Paralympics in history, much of the support given to those with a physical or mental impairment was being taken away. There is a sense as we complete the first quarter of the 21st century, the access of goods, services and employment to those with a disability is not a right that is granted.

The Equality Act 2010

The current legal framework concerned with the provision of accessibility to goods, services and employment is the Equality Act 2010 (EA10). This legislation is applicable throughout Great Britain, although the devolved powers in Scotland and Wales permit minor alterations to the text. As discussed, the EA10 replaced the Disability Discrimination Act 2005 (DDA05), which in itself was a revised version of the Disability Discrimination Act 1995 (DDA95). The EA10 is intended to prevent discrimination to a much broader sector of society than the DDA05 and includes a list of 'protected characteristics' including age, gender reassignment, being married or in a civil partnership, being pregnant or on maternity leave, disability, race (including colour, race, nationality including ethnic or national origin), religion or belief, sex and sexual orientation.

There has been some open criticism of the decision to repeal the DDA05 and absorb this in the EA10 with the suggestion that bringing the Disability Rights Commission into the Equality Rights Commission has resulted in a loss of focus on disability issues.[5] It has also been suggested that it would have been better to retain the DDA05 rather than absorb this into the EA10.[6] Stronger criticism of the introduction of the EA10 has been levied by UNISON as this new legislation is seen as *weaker than the previous [legislation]*. Specifically cited is the effect of diluting the protection offered in the original 1995 Disability Discrimination Act (DDA95). There is much historical discussion evident to suggest that the EA10 has diluted some of the initial gains made by the DDA95. This is evident with the replacement of the term in the DDA95 (referring to discrimination) as 'less favourable' with the more general term of 'unfavourable' in the EA10. Such fine nuances may be exposed as and when the EA10 is 'tested' in a court of law.

The Equality Act 2010, as expected, is a much larger document[7] than the DDA95 and in essence is divided into the following 16 parts:

1   Socio-economic Inequalities;
2   Equality: Key Concepts;
3   Services and Public Functions;
4   Premises;
5   Work;
6   Education;
7   Associations;

8 Prohibited Conduct: Ancillary;
9 Enforcement;
10 Contracts, etc;
11 Advancement of Equality;
12 Disabled Persons: Transport;
13 Disability; Miscellaneous;
14 General Exceptions;
15 Family Property and
16 General and Miscellaneous.

The Equality Act Part 2 – Equality: Key Concepts

In the context of providing access to goods, services and employment, Part 2 is the first section of the act of significance. In particular this refers to what are known as 'protected characteristics', of which disability has its own definition. In essence the definition of disability is if a:

* *person has a physical or mental impairment and*
* *the impairment has a substantial and long-term adverse effect on a person's ability to carry out normal day-to-day activities.*

In context, many individuals may not consider themselves to be 'disabled', and a good example of this might be the elderly. A staunch or stoic approach to getting on with the demands of daily life when struggling with mobility issues would certainly not engender some of pension age to consider themselves as 'disabled'. However, when asked if they have a long-term physical or impairment affecting their ability to carry out normal day-to-day activities, many more resonate with this idea. Accordingly, this is why a large percentage of those above the age of 65 are identified in published government data as being disabled, and this has been extensively discussed in Chapter 3. Guidance has been published by the UK Government on the terms what is considered to be:

* Substantial.
* Long Term.

The term *substantial is more than minor or trivial, e.g., it takes much longer than it usually would to complete a daily task like getting dressed* and long term equates to *12 months or more, e.g., a breathing condition that develops as a result of a lung infection.*[8] The definition of disability also covers progressive medical conditions such as cancer, HIV infection and multiple scoliosis. There is further government guidance on recognised conditions, and those not considered to be a disability have been published by the UK government.[9] There are a number of case studies or examples included in the guidance, and the nuances of disability or impairment are considered in the publication. For example, obesity is not considered as an impairment; however, subsequent difficulties in breathing or mobility could adversely affect the ability to walk, therefore the guidance indicates it is *the effects of these impairments that need to be considered, rather than the underlying conditions themselves.*

Although this is intended to be guidance, it does appear to recognise there are 'grey areas' and, as established with the disabilities listed in Chapter 3, there are a number of umbrella terms for some conditions. These terms and also the understanding that individuals with specific conditions can have a range of mild to serious symptoms means that the definition of disability should relate more to the ability to undertake normal day-to-day activities.

Included in the government guidance is a specific list of conditions not considered to be a disability and these include:

- *addiction to, or dependency on, alcohol, nicotine, or any other substance (other than in consequence of the substance being medically prescribed);*
- *the condition known as seasonal allergic rhinitis (e.g., hay fever), except where it aggravates the effect of another condition;*
- *tendency to set fires;*
- *tendency to steal;*
- *tendency to physical or sexual abuse of other persons;*
- *exhibitionism and*
- *voyeurism.*

Part 2 of the EA10 also details *Prohibited Conduct,* and this further defines discrimination across the spectrum of all protected characteristics. In the context of discrimination against those with a disability, the Act indicates that someone cannot be treated unfavourably because of this status. Furthermore, the perpetrator is deemed to discriminate if they cannot show the treatment of those with disability as proportionate to achieving a legitimate aim. While this appears to make sense when accessing goods, services and employment, there is an exception to it. If it is not possible for the supplier of goods and services or the employer cannot know or reasonably be expected to know of the disability, then they cannot be deemed to have discriminated. This is important in the context that not all disabilities are visible, which may not be realised in a one-off visit to a commercial property. It is more prevalent with employment, which is a longer-term engagement between organisations and an individual. Those with disability are not obliged to disclose anything; however, in the context of then proving any subsequent discrimination, this will be difficult.

Also covered in Part 2 of the EA10 is the notion of 'reasonable adjustment', and this is something that will continually be referred to in this book. Section 20 of the Act refers explicitly to 'duty to make adjustments'. This comprises of three fundamental requirements:

- Take reasonable steps to avoid the disadvantage of disability (discrimination).
- Take reasonable steps to avoid the disadvantage of disability (discrimination) in the context of the presence of a physical barrier.
- Provide an auxiliary aid to prevent the disadvantage of disability (discrimination).

An important principle in the legislation is that those with a physical or mental impairment are not expected to pay the service provider any costs for taking the reasonable steps. Furthermore, where there is a physical feature engendering a disadvantage the Act advocates:

- removal of the feature;
- alteration of the feature or
- provide a reasonable means of avoiding the feature.

A physical feature by definition in the Act is:

- something arising from the design or construction of a building;
- a feature on the approach, exit and access to a building;
- fixtures, fittings or equipment in or on a premises and
- any other physical element or quality.

The Equality Act Part 3 – Services and Public Functions

Part 3 of the EA 10 concerns the service providers, and while it also covers public service provision and is intended to prevent harassment as well as victimisation, Part 3 does explicitly cover discrimination on the basis of providing a service. It also obliges the need to make reasonable adjustment from a service provision perspective. Part 3 is not explicitly associated with the physical access of premises. In essence, this relates to the operation of the service, which can be significantly different to that of the building. For example, there are a number of access components to be considered when facilitating access for a blind person to a cinema, and this is covered by Part 4 of the act. Accordingly, Part 3 will relate to the cinema experience and for a blind person will include the need for the film to include an audio description. This book concerns primarily the physical access to goods, services and employment.

The Equality Act Part 4 – Premises

This book is concerned with accessibility and inclusive environments; therefore, the prime focus is the provision of access to goods and services for those with disability. As observed; the Equality Act 2010 covers a wide range of protected characteristics, and the generalisation of this legislation is 'exposed' when contextualised with the section on premises. It is clear that this book is seeking to detail disability-specific nuances of the relevant legislation. However, there is no specific definition of 'premises' within Part 4 of the Act. The Act places responsibilities on those disposing or managing premises. In respect of providing reasonable adjustment, Section 36 of the Act places an obligation on the landlord or tenant of let premises or the person managing the common parts to facilitate this. In essence, while premises are defined as those owned, leased or managed, it is Part 2 of the Act detailing the key concepts and particularly in Section 20 which covered the notion of providing reasonable adjustment.

The Equality Act Part 5 – Work

While Part 5 covers 'work', it does not explicitly detail work and disability; the default position for the application of the Act and disability in the work environment concerns that defined in Part 2. There is a subsequent *Statutory Code of Practice Employment* published in 2011 by the Equality and Human Rights Commission.[10] This document will be discussed later in this chapter under the subheading guidance; however, it is excellent in detailed the application of the Equality Act 2010 within the work environment.

The Equality Act Part 6 – Education

In line with the other specific parts of the Act concerning premises and work, the part relating to education discusses in general terms the notion of discrimination and the requirement to provide reasonable adjustment. There is a focus on the service provision or delivery of education and less so the physical access to educational establishments.

The Equality Act Part 9 – Enforcement

Breaches of the Equality Act 2010 are addressed through civil procedure, employment tribunal or judicial review. Effectively a successful claim under the EA10 may result in the payment of damages.

While the EA10 does not explicitly list examples of reasonable adjustment, this is likely to be shaped by legal challenge. Since its enactment, there have been a number of publicised claims made under the legislation concerning the protected characteristics such as race, gender and sexual orientation; these have been made in the context of discrimination in the workplace on gender inequality, as well as in the public domain in the delivery of goods or services. An online search of news sites quickly reveals a number of challenges under the Equality Act 2010, but there appear relatively few associated with the physical access of commercial property for access to goods, services or employment.

The highest profile case concerning the EA10 is a legal challenge mounted in *Paulley v First Group plc*. This case has been widely reported in the national printed and online press and is also effectively summarised by Disability Rights UK[11] as it went all that way to the Supreme Court. In essence, Doug Paulley, a wheelchair user, tried to board a bus, but the wheelchair space was occupied by a mother with a sleeping baby in a pushchair. The driver asked the woman to fold down the pushchair, but she said this was not possible. According to the publicly accessible judgement of the Supreme Court,[12] Mr Paulley offered to fold his own wheelchair and occupy an ordinary passenger seat. This offer was refused by the driver on the grounds that there was no safe way to secure the folded wheelchair on what would be a journey on a rather winding route. Mr Paulley could therefore not take the bus and had to wait for the next bus, which meant missing a later train connection. After initially winning damages, First Group PLC appealed the decision, which eventually escalated to the Supreme Court, which upheld the claim.

The case was seen as a landmark decision, but there are observers critical of the decision in the sense that it was argued that it is not possible for a driver to force someone to give up the wheelchair space, if inappropriately occupied. Furthermore, on a full bus and asking someone to actually vacate a bus mid journey was considered unreasonable, particularly when they are already lawfully on the bus. It was also successfully argued that an elderly person occupying the space or a disabled non-wheelchair user (with possible mobility issues) could not be expected to give this up. In summary, there is a legal provision afforded by the Equality Act 2010, but while damages were awarded, it has not been seen in this case to wholly prioritise disability in this matter. Further criticism on the obligation of individual disabled complainants was reported by

Disability News Service (2020)[13] marking 25 years of disability discrimination legislation. This article concluded that a lack of legal aid, as well as the high cost of litigation, prevents claims and is doomed to fail. This opinion is from those representing disability suggests that there is an element of box ticking on policies such as diversity and inclusion. What is the point of having a policy or even a wheelchair space on a bus if it cannot be prioritised for wheelchair users, irrespective of legal protection? The commercial world appears to have adopted a check box approach to legal provision or minimal legal standards, and this is appearing evidently widespread to commercial property.

Within the built environment it can be commented that while there are a wide range of legal requirements for the construction, operation and occupation of commercial buildings, there appears a willingness to pay 'lip service' to this idea. Put frankly, many of these requirements are seen as onerous and costly, with little benefit. Of course, there are certain legal requirements, such as fire safety or even health and safety, that can have graphic, high-profile exposure in the event of failure. As these are life safety issues and have the potential for criminal proceedings as well as reputational damage, they seem to draw more attention than those requirements such as disabled access, which appears to be perceived as onerous. Chapter 7 discusses both public and commercial attitudes to disability in the built environment; it challenges the stick-over-carrot approach to providing an inclusive environment.

Equality Act 2010 Statutory Code of Practice Employment

Published in 2011 by the Equality and Human Rights Commission, the *Equality Act 2010 Statutory Code of Practice Employment*[14] provides clear guidance on the application of the EA10 in a work environment. It is a comprehensive document that comprises 19 chapters plus appendices covering all aspects of the Act. This guidance document covers all of the protected characteristics within the chapters and sub-chapters; disability is initially addressed in Chapter 5; *Discrimination Arising from Disability*. There are a number of worked examples or case studies illustrating and defining how discrimination can occur in the workplace. It guides employers on how to reduce the possibility of discrimination occurring and the relevant steps to mitigate the risk of this. Chapter 6 is probably the most relevant in the context of making reasonable adjustment for employees with a physical or mental impairment. In essence, there are three principles to making a reasonable adjustment for disability in the workplace:

- Avoid substantial disadvantage in all aspects of the workplace or work practice for those with a disability when compared to non-disabled employees.
- Remove or alter a physical feature or provide a reasonable means of avoiding a physical feature which puts a disabled person at a substantial disadvantage when compared to a non-disabled person.
- Provide an auxiliary aid or auxiliary service to those with a disability to prevent any substantial disadvantage when compared to a non-disabled person.

The first point concerning avoiding 'substantial disadvantage' is concerned with all aspects of the workplace and work practice for those with disability. It appears wide open and ambiguous, but in essence, the code of practice states that significant disadvantage of those with disability compared to those non-disabled includes an organisation's:

- formal or informal policies;
- rules;
- practices;
- arrangements or
- qualifications including one of decisions and arrangements.

Physical features are defined as being either permanent or even temporary and moveable items such as furniture but mostly include part of a commercial property's design, construction or operation such as:

- parking, external paving surfaces, steps and kerbs;
- building entrances exits, including those for emergency use;
- internal areas including doors and floor surfaces as well as signage and furniture;
- toilets and
- lifts, lighting and ventilation.

Examples of physical features associated with impeding access to the workplace or in the delivery of goods and services within the built environment are not illustrated in the code of practice. However, these will be discussed in much more detail in Chapter 9 of this book, including examples of the legal minimum requirements and best practice.

Concerning the concept of providing an auxiliary aid to prevent disability discrimination, this is likely to be far more bespoke or individualised to the relevant member of the workforce. The requirement should be tailored to the needs of the relevant disability requirement. In many respects, this is similar to the 'service' part of the service provider, while the actual physical access to a building or workplace is largely black and white, in the sense that access can or cannot be permitted; there is more nuance to auxiliary aids. This could include specialist equipment such as adapted keyboards, hearing loops or even a sign language interpreter or support worker for someone with a disability. The disclosure of an employee's disability will require the analysis of any disability-specific needs in respect of the provision of auxiliary aids by an organisations human resource department or management.

The provision for ensuring the prevention disability discrimination does largely rely on the disclosure of an employee to enable an assessment to be made of the specific disability needs. Under the legislation (EA10), an employer is required to make reasonable adjustment if they know or could be reasonably expected to know that a worker is disabled. This may require a confidential objective assessment which maintains the privacy and dignity of the employee.

Reasonable Adjustment

The term 'Reasonable Adjustment' appears highly ambiguous in the sense that there is no definition of 'Reasonable', which is very much a term used in legal language. In the context of providing access to commercial property, goods, services or employment, surely a reasonable adjustment should be evaluated from both the service provider's and service users' perspective. It appears misbalanced or unequal by virtue that currently the notion of reasonable adjustment is assessed from the perspective of the service provider or employer. The *Equality Act 2010 Statutory Code of Practice Employment* analyses what constitutes a 'reasonable adjustment' in the workplace. Is suggests that this can take the form of physical alteration to the work premises, such as the widening of doors, placement of ramps, or even the installation of an accessible toilet. It should be noted that these can have significant cost implications. However, reasonable adjustment can also take the form of the provision of auxiliary aids, as previously discussed, or adjustment to the working practices of those with a physical or mental impairment. There clearly appears to be an obligation placed upon the employer or service provider, who may not be disabled and have an experiential understanding of the relevant access challenges and what is actually reasonable. The notion of 'experiential understanding' is discussed in Chapter 7 through analysis of research published into public and commercial attitudes to disability in the Built Environment. In essence, it concludes that only those with a physical or mental impairment or those caring for someone with a physical or mental impairment can understand the real access challenges directly through their experience of it.

With respect to undertaking physical building alterations, one consideration is that if a premises is leased, then it may be necessary for the employer to request a license to alter for any physical changes to the property.

Under the Equality Act 2010 and where there is an obligation placed on the employer in their capacity as a tenant to request consent to make a physical alteration to prevent discrimination, it is deemed not reasonable to do this prior to receiving landlord approval. Furthermore, the landlord cannot withhold consent reasonably; however, it is recognised in the legislation that if such alteration resulted in a substantial and permanent reduction in the value of their premises or interest, then this can be withheld. Additionally, if an alteration were to significantly disrupt or inconvenience other tenants in a multi-let property or site, it could also be withheld. Where there is an obligation to comply with other statutory requirements such as planning, building regulations, listed building consent or fire regulations, the EA10 does not overrule these. If a proposed reasonable adjustment introduces a health and safety risk to the building users and this includes the person(s) with the disability, then this may be a relevant deciding factor. There will be a need to undertake sufficient risk assessment to determine this likelihood and consequence of any risks.

Factors affecting the implementation of a reasonable adjustment may also include:

- the overall effectiveness of this;
- the practicality of implementation;
- financial, other costs and disruption on implementation;
- employers available financial or other resource;
- availability of financial or other assistance to make adjustment and
- the type and size of employer.

It appears striking that the definition of a 'reasonable adjustment' is only considered from the perspective of the employer or commercial organisation. As this is deemed as the reasonable steps to be taken commercially, then both the incentive and control of access are out of the hands of someone with a physical or mental impairment. In essence, along with Mike Oliver's *Social Model* of disability, there appears a *Commercial Model* of disability too. There is a need to consider reasonable adjustment from the perspective of those with a physical or mental impairment. This is complicated by the variety of disabilities and the diverse, nuanced requirements of each physical or mental impairment. Based upon research published in 2024 and illustrated in Chapter 3, it can be noted the access priorities of each listed category of disability. This goes some way to helping establish what might be reasonable from the perspective of disabled users and reinforces the three-dimensional nature of accessibility. It is not one solution for one disability but more disability-specific solutions across a range of access variables; despite this, it has been established in Chapter 3 that it is common for those with a disability to have more than one physical or mental impairment. This appears influential and logical in the historic adoption of 'blanket', one-stop strategies to cover all bases with respect to the different types of disability. Add to this the complexity of implementing reasonable adjustment in the form of access alteration to existing buildings, and it is easy to see why there has to be a degree of compromise. However, it appears unjust to look at compromise from purely a commercial perspective; good access solutions will need to consider the perspective of those with a physical or mental impairment.

In the context of implementing reasonable adjustment from a working practices perspective for those with disability, the *Equality Act 2010 Statutory Code of Practice Employment* identifies the following examples:

- undertaking adjustment to the premises (as discussed previously);
- providing information in accessible formats such as Braille, audio tape, audio description or publications in 'Easy Read' format, this could include employment or recruitment information;
- reallocation of work duties between disabled and abled bodied employees;
- transferring a disabled worker into a more suited existing role;
- permitting flexible working, extra breaks, shorter shifts or working from home for disabled employees;
- assigning disabled employees to a more suitable place or building of work;
- allowing disabled employees time for recovery or treatment;
- giving or arranging training or mentoring for disabled or other employees;
- acquiring or modifying equipment;
- modifying procedures for testing or assessment associated with employment or recruitment;
- providing a reader or interpreter;
- providing supervision or support to disabled and able-bodied employees;
- permitting disability leave for rehabilitation to return to work;
- employing a support worker to assist a disabled employee;
- modifying employment procedures for disabled employees including disciplinary, grievance, and redundancy policies and
- modifying performance related pay arrangement for disabled employees.

Making reasonable adjustment to working practices and policies may be significantly more cost-effective compared to some physical alterations to buildings, and the option to adopt procedures such as relocating disabled staff may also be seen as a reasonable adjustment in itself.

Building Regulations – Approved Document Part M

Building Regulations comprise a series of 'Approved Documents' relating to all aspects of construction and are issued by the relevant secretary for HM government, as well as separate versions from the governments of the devolved nations. Approved Document (AD) Part M concerns *Access to and Use of Buildings*. It is divided into two volumes with *Volume 2 – Buildings Other Than Dwellings* applying to commercial property. The current version of AD Part M is associated with Building Regulations 2010, and currently, there is the 2015 Edition that has incorporated amendments in 2020.[15]

Seen by those in the construction sector as prescribing to the legal minimum standards for both new build properties as well as those the subject of significant renovation, Part M is perceived as being the default requirement for accessibility. However, in the preface of the document, the text makes specific reference to the Approved Documents providing guidance for the more common building situations. It recognises that there may be more than one way to achieve compliance, but importantly, the AD Part M does state that its adoption is not obligatory. There is an explicit reference to the Equality Act 2010 in AD Part M, with a statement that compliance with Building Regulations does not necessarily equate to compliance with the EA10. This is with the understanding that the EA10 obliges service providers and employers to make reasonable adjustment to any physical feature that may disadvantage a disabled person.

In most normal circumstances, building regulations are applicable to new build projects as well as those the subject of renovation. A significant quantity of commercial properties exists, and Part M is applicable to renovation or the structural alteration of these, as well as any alterations to the entrances, exits or fire safety. Even a material change of use means that Part M is applicable regarding access. However, implementing reasonable adjustment within the existing built environment will inevitably result in compromise; it cannot be underestimated the complexity of applying current regulations retrospectively to existing buildings. With a large quantity of commercial property having been built prior to the current building regulations, it is necessary to consider the previously discussed notion of 'reasonable adjustment'. It has been established with the *Equality Act 2010 Statutory Code of Practice Employment* what reasonable adjustment or reasonable steps are likely to mean to employers and the removal or alteration of physical features. It has been observed that by presenting an alternative solution to eliminate the barrier to access, then a reasonable adjustment to service provision can be achieved. Despite this, there is no definition or notion of what is 'reasonable' from a user perspective. Therefore, it is necessary to default to the legal minimum requirement established in the AD Part M and contextualise this with a bespoke assessment of physical features that may be present to a commercial property or place of work. There are standards considered to be above the legal minimum for inclusive design, and these are primarily British Standards, which will be discussed later.

Approved Document Part M Volume 2 is divided into the following principal sections:

- Section 1: Building Approach Route and Parking

  - Level Approach
  - Onsite Parking and Setting Down
  - Ramped Access
  - Stepped Access

- Handrail (Steps and Ramps)
- Hazards on Access Routes

- Section 2: Principal Entrance of Main Staff Entrance

  - Accessible Entrances
  - Doors to Accessible Entrances
  - Manually operated and Non-Powered Entrance Doors
  - Powered Entrance Doors
  - Glass Doors and Glazed Screens
  - Entrance Lobbies

- Section 3: Horizontal and Vertical Circulation

  - Entrance Hall and Reception Area
  - Internal Doors
  - Corridors and Passageways
  - Internal Lobbies
  - Provision of Lifting Devices
  - Passenger Lifts
  - Lifting Platforms
  - Wheelchair Platform Stairlifts
  - Internal Stairs
  - Internal Ramps
  - Handrails to internal Steps, Stairs and Ramps

- Section 4: Facilities

  - Audience and Spectator Facilities
  - Refreshment Facilities
  - Sleeping Accommodation
  - Switches, Outlets and Controls
  - Aids to Communication

- Section 5: Sanitary Accommodation

  - Sanitary Accommodation Generally
  - Provision of Toilet Accommodation
  - Wheelchair-accessible Unisex Toilets
  - Toilets in Separate Sex Washrooms
  - Wheelchair-accessible Changing and Showering Facilities
  - Wheelchair-accessible Bathrooms

The Building Regulations Approved Document Part M is available online as a free-to-access document published by the UK government.[16] The document is highly technical and gives both descriptive text as well as illustrations on the design or audit of disabled access provision. Chapter 9 of this book will look in more detail at the implementation of accessible features within the built environment and the practical application. The general concepts for each section and subsection of Approved Document Part M are summarised as follows:

Section 1: Building Approach Route and Parking

In essence, the building approach does not constitute the local authority approach route such as the streets or pavements owned and maintained by the local authorities. However, much of the principles detailed in Section 1 could be applied to the adjoining streets and pavements. Essentially the building approach begins at the entry point on a boundary or between different buildings on the same privately owned site. It should be recognised that changes in level are probably one of the most common challenges to those with a disability. If the site within the boundary is a completely new development and essentially a 'blank canvas' for designers, it may be easier to adopt a level approach from the concept. However, as most of buildings and site access relate to the retrospective (post-1995) provision of access, different levels are often quite common. Furthermore, the natural environment is rarely flat and level; the locations of most commercial properties offering goods, services or providing places of employment will have a degree of level change to navigate. Allied to a level approach, access routes should also be sufficiently wide enough to accommodate both wheelchair users, the disabled walking with sticks or other aids and the non-disabled. The width has defined criteria in Approved Document Part M, but the principle should aways be to allow sufficient provision for the passing of both disabled and non-disabled persons on the same access route.

The building regulations give specific details on the following aspects of the building approach and parking:

- changes in level;
- width of access route including the provision of passing places;
- gradient and 'landing' levels;
- removal of trip hazards and obstructions;
- surface texture (firm, durable and slip-resistant);
- placement of hazard warning surface ('blister paving');
- clearly identifiable and well-lit main entrance;
- appropriate parking provision for disable persons arriving by car with space to transfer into a wheelchair and manoeuvre barrier-free on to the access route;
- ramped access (gradient, landings, handrails, signage, surface texture [including colour contrast between landings and ramp], stepped alternative);
- stepped access (tactile paving at top and bottom of steps, tread material, coloured nosings, riser detail, landing dimensions);
- handrails to steps and ramps (vertical heights, provision for continuous handrails, end detail/extension, visual contrast, slip resistance, thermal conductivity, profile, positioning) and
- hazards on access routes (obstructions along and above).

Section 2: Principal Entrance of Main Staff Entrance

In essence, the Building Regulations Approved Document Part M states that for new buildings, the principal entrance or main entrance for staff and entrance lobby should be fully accessible. Furthermore, the document recognises that for existing buildings that were designed and built pre-Part M, the principal entrance may not be accessible.

Accordingly, the document advocates the provision of an accessible alternative. As discussed in Chapter 5, the types and characteristics of commercial buildings are diverse. Many high streets date pre-1920 and have an array of unique, historic buildings operating as shops or small businesses. These often do not have a secondary entrance, and therefore, the main entrance is the sole entrance; certainly the necessary alteration or adaption of these entrances to provide an inclusive environment may result in 'reasonable adjustment'. Such adjustment is potentially further complicated if the building has historical or cultural value and is either listed or placed within a *Conservation Area*. This situation is fully explored in Chapter 8, but essentially, there will be a need to comply with both the Equality Act 2010 and planning legislation.

The AD Part M stipulates significant design details concerning the accessibility of the main entrance, but there are some key principles that need to be addressed. Where the approach to the principal entrance is steeply sloped or access is restricted, then it is necessary to foresee an alternative entrance. This will of course be dependent upon the location of the commercial property, and the nature of many town or city centres is that these are rarely full flat (Fig 4.1).

The document also stipulates that the entrance should be well signposted or easily recognisable, and while recognising that door thresholds should provide weather resistance, they should also not present a trip hazard. The provision of signage is something typical of commercial office buildings or purpose-built shopping centres. The condensed nature of town centre retail units or bars, cafes and pubs means there is often only one entrance, and this is often clearly recognisable due to the use class of the property. Where there may be the provision of an accessible entrance as an alternative, this should be explicitly signposted (Fig 4.2).

Looking more explicitly at the technical detail concerning the individual components that contribute to the principal or main entrance, AD Part M sub-divides these into the following:

Figure 4.1 (a) Above left: A steep shopping street in the commercial heart of a city. (b) (right) the challenging topography makes step free access difficult to achieve.

Figure 4.2 (a) Above left: Circled is the inaccessible stepped entrance to a pre-1920s office and marked in the rectangle is the alternative step free access. (b) Above right: A modern post-2010 office building with inexcusably and inaccessible principal entrance. (c) Marked on the glass (right) is information identifying the accessible entrance which is barely visible from the street.

- Accessible entrances are clearly signposted with internationally recognised symbols, well-lit or encompassing colour contrast between the walls and doors, the entrance should be free from obstruction and have an adequate landing space in front of the door. Any door-entry intercom systems should be accessible and surfaces, including doormats, should not present a trip hazard.
- Doors to accessible entrances should be wide enough to facilitate wheelchair access and can be manual, power-assisted or fully automatic. Overhead closing mechanisms of manual doors may prohibit those with a physical impairment from access and vision panels should be placed in the doors to avoid collisions.
- Automatic power assisted should have sufficient activation through motion sensors or push buttons to allow sufficient time lapse for disabled users. The doors should incorporate sufficient safety features and default to manual operation in the event of fault or power failure. The position of push buttons and controls should be at accessible heights.
- Glazed doors, panels and screens should have etching of manifestation on the glass to ensure those with a visual impairment are in no doubt about the presence of the glass.

- Entrance lobbies should in essence be large enough for a wheelchair user and companion or carer with sufficient space for the closure of one door before opening the other. The floor surface should not impede access or present a trip hazard.

Section 3: Horizontal and Vertical Circulation

The concept of horizontal and vertical circulation relates to the use occupation of the building; it stipulates that the purpose of providing access should be for disabled users to have access to all relevant facilities. While not explicitly indicated in the Building Regulations Approved Document Part M, vertical circulation is considered to be one of the most significant obstacles for those with a physical impairment. It is also one of the costliest to accommodate if the design solution is to provide vertical transport such as a lift. Therefore, a reasonable adjustment may be the provision of the exact same goods, services and employment available on multi-levels of a building without having to transfer between levels. This may be evident in a two-storey restaurant where the same service is replicated on both levels. However, this may not be the case if the said restaurant markets itself as say having unique dining with excellent first-floor views; this might be the case in the marketing of a '*Sky Bar*'. Clearly not being able to provide access to these defined and specific features would be discrimination.

In the context of providing both horizontal and vertical circulation, the Building Regulations Approved Document Part M looks specifically at the following:

- Entrance halls and reception areas should be located in sight of the main entrance with a clear, unobstructed access route but sufficiently away from noise generated at the entrance. The floor surface should be non-slip and any reception desk should be sufficient for those in a wheelchair, sitting or standing. There is a requirement to provide an induction or hearing loop.
- Internal doors potentially pose an obstruction and where these have closing mechanisms the force required to open these should be limited, doors should have sufficient clearance with accessible door latches or handles. The doors, door frames and door furniture should be in contrasting colours to aid those with visual impairment. Glazing and vision panels to doors should be uniform in dimension and placement as well as utilising etching or manifestation. Fire doors should use electromagnetic devices to hold these open and operable with activation of smoke detection or fire alarms.
- Corridors and passageways should have sufficient width with no protruding obstructions such as radiators, and where these exist, they should contrast visually in colour and have a protected guardrail. Floor levels, gradients and slopes should be controlled with landings as required. Doors opening into a corridor should be recessed to avoid constituting an obstruction, doors located in corridors should comply with internal door requirements including vision panels or glazing. Floor finishes should be non-slip, and patterned flooring should be avoided as this can be mistaken as steps or changes in level.
- Internal lobbies should have minimal length and width provision which also accommodates sufficient space between swing doors. Floor finishes and the presence of glazing panels should not present a trip hazard and reflective distraction, respectively.

- Provision of lifting devices is probably one of the most complex and costly access provisions. All 'new build' properties should have a compliant lift serving all storeys, and contrasting signage should be present indicating the presence of lifts as well as the relevant floor-level indication. There should always be the provision of stairs additional to any lift installation, and ramps can also be installed as an alternative. The general requirements for the provision of lifting devices and landings should sufficiently illuminated to avoid glare or shadows and the call buttons or controls should be reachable by those in wheelchairs, colour contrasted, and have a tactile or braille finish. Floor surfaces of lifting devices should be frictionless with handrails and emergency communication provided.

  - Passenger lifts are generally standardised to comply with regulations, standards and norms. Accordingly, there is a requirement for sufficient door width, lift car size, audio and visual signalling, as well as sufficient door opening or closure for disabled users.
  - Minimum lift dimensions are stated in building regulations approved document Part M as are some of the technical details and the reasoning for these. This includes the time-lapse for door closure as well as audio and visual signalisation, where the minimum lift dimensions do not permit a wheelchair to turn 180° a mirror should be installed.

- Lifting platforms have similar requirements to lifts but one key feature is that these are operated by continually pressing on the control button for the duration of the vertical journey. They can be 'open' and non-enclosed up to a certain height dictated in the regulations; thereafter, they need to be enclosed in a shaft or something similar.
- Wheelchair platform stairlifts are considered to be the lowest level or vertical transport system and should only be installed when it is not possible to place a lift or lifting platform. These are often used as a compromise to providing access within an existing building where it is not possible to place a lift, lift shaft or platform lift, they may be found in listed or historically important buildings. One key requirement of these installations placed in staircases is that they do not adversely affect the means of escape from the building.
- Internal ramps, staircases and handrails adopt largely the same design criteria as for those placed externally facilitating access to a building. These are also required to comply with Building Regulation Approved Document Part K: *Protection from Falling, Collision and Impact.*

Section 4: Facilities

The provision of an inclusive environment primarily concerns the physical access to goods, services and the places of employment. The accessibility of the goods themselves is something different, and due to the diverse nature of human existence and society as a whole, it is not possible to micro analyse the nature of every service provider. An example of this might be with the access provision to a cinema, and even with relatively complex buildings, this is possible. However, it does not mean that the film is accessible to blind people or those with visual impairment. In such circumstances, there is a need for the cinema to provide the facility for audio description. There needs to be a dual approach to providing access to goods, services and employment; for the disabled end user or employee, it serves no benefit if a building is accessible but the prime service is not and *vice versa*.

Accessing facilities may be considered as the prime focus of delivering an inclusive environment, but with the logical sequencing of building regulations, this is one of the last areas of concern. There is a recognition that to access facilities, it is necessary to facilitate the main access routes, parking (if applicable), main entrance as well as horizontal and vertical circulation. Building Regulations Approved Document Part M considers 'facilities' with the same level of detail as other sections of the document addressing access. The following is a summary of the key points detailed within the regulations:

- Audience and Spectator Facilities are primarily considered to be lecture spaces or conference facilities, entertainment facilities such as theatres or cinemas, or sports facilities such as stadia. The general design considerations for users of these facilities include the need for users to view or listen from one side, and this includes being able to lip read or read sign interpreters. Venues should be accessible for wheelchairs, and seating provision should allow for adjacent seating for carers or companions. Accessible seating should foresee rest space for assistance dogs, and the venue can consider a number of removeable seats to facilitate a greater number of accessible seatings, above and beyond the minimum stipulated in the regulations or even to facilitate extra leg space. For lecture spaces and conference facilities, there is as requirement that all participants can use the presentation facilities. Approved Document Part M clearly prescribes the necessary requirements and these include:

  - access routes to wheelchair spaces;
  - compliance with the technical requirements for steps and staircases;
  - stipulated minimum requirements for permanent wheelchair spaces as well as removeable seating;
  - the possibility for a range of viewing points where multiple spaces are provided;
  - level or horizontal spaces;
  - fully accessible podium or stage for lecture or conference facilities and
  - hearing enhancement systems (hearing loops).

- Refreshment Facilities is an umbrella term to describe restaurants and bars, the building regulations stipulate that these should be designed to ensure they can be accessed by all. For those with disability, these should be accessible independently or with companions. Accessibility should also concern staff working at these facilities. The general design considerations include the access to public areas, toilets and external terraces. It is acknowledged that refreshment facilities may be self service or waiter service and Approved Document Part M stipulates that users should have access to both. As with the provision of internal circulation, internal changes in level between different zones or areas in refreshment facilities should foresee the placement of a ramp. The necessary design provision concerns:

  - access to all parts of the facility;
  - counter heights;
  - worktop heights and
  - wheelchair accessible thresholds.

- Sleeping Accommodation from a commercial perspective and as defined in Approved Document Part M includes hotels, motels and student accommodation. The practical application of this would also include Bed and Breakfast accommodation and increasingly could affect a number of additional leisure facilities where there is sleeping accommodation. There is a need to consider the size of an establishment, and in the terms of the application of the building regulations, the requirement is for 1 in 20 bedrooms to be accessible. In reality, this contradicts the obligations of a service provider under the Equality Act 2010, which obliges those providing sleeping accommodation not to discriminate on the basis of disability. This no doubt could be legally tested on what constitutes reasonable adjustment; however, there is also no doubt that there is commercial advantage in providing a fully accessible bedroom, irrespective of an establishment having fewer than 20 rooms. Approved Document Part M addresses sleeping accommodation with a list of requirements concerning:

  - the location of accessible rooms;
  - considerations for power-assisted entrance doors and door widths;
  - space to transfer from a wheelchair to the bed unassisted;
  - preference for ensuite sanitary facilities and compliance with the necessary dimensions;
  - the ability to visit companions in other rooms with accessible entry doors;
  - the design criteria for fitted wardrobe doors and handles including colour contrast;
  - window opening mechanisms;
  - accessing balconies and
  - the provision of visual fire alarms as well as emergency assistance call points.

- Switches, Outlets and Controls should be visible and therefore require contrasting colours against their background. They need to be placed at appropriate heights, as detailed in the regulations, and need to consider touch plate controls for visually impaired people as well as those with limited dexterity. There is an obligation to place outlets and controls in uniform locations at specific heights.
- Aids to Communication concerns way findings as well as public address, and while the principle of colour contrast between walls, floors and ceilings, there is a need to consider electronic aids to communication. This includes visual information supplementary to a clear and audible public address system in addition to hearing enhancement systems, such as induction loops, telephones for hearing aid users and text telephones. For certain aids to communication, including the Approved Document Part M, defaults to the guidance are available in British Standard (BS) 8300.

Section 5: Sanitary Accommodation

Perhaps along with the presence of access ramps for building entry, accessible sanitary rooms are perceived as a marked attempt to create an inclusive environment. These are considered to be a normal occurrence across the range of commercial buildings detailed in Chapter 5 of this book. Despite this, there are a variety of different accessible sanitary facilities that have their own appropriate technical requirements. The principal design consideration for these facilities is a need to recognise the existence of different disabilities and how to accommodate them. In essence, there is a requirement to ensure that taps are automatic or can be operated with a closed fist, and they need to comply with specific requirements for water fittings. Cubicle doors, handles and locking mechanisms should have minimal set resistance with doors opening outward. In the event of someone collapsing in the cubicle, there should be appropriate emergency alarm facility as well as a means to access the door externally including release of locking mechanisms. There is a need to ensure adequate lighting as well as ensuring heat emitters and the operating temperature do not pose a risk or scalding. The principle of contrasting finishes should be applied to sanitary fittings, grab bars with wall and floor finishes. In essence Approved Document Part M of the Building Regulations contains significant technical detail and addresses the following fundamental requirements for sanitary accommodation.

- Provision of Toilet Accommodation as stipulated in the regulations include:
  - cubicle types;
  - enlarged cubicles;
  - placement of a wheelchair accessible unisex toilet is a minimum requirement where only one toilet is provided;
  - the placement of accessible cubicles where ever there is sanitary accommodation for building users;
  - the placement of sanitary equipment (toilet and toilet seat, wash hand basin, hand drier, grabrails etc including emergency assistance call-cord);
  - the provision of WC cubicles for ambulant disabled people and
  - the provision of changing places to assembly, recreation and entertainment buildings.

- Wheelchair-accessible Unisex Toilets details the following:
  - location and position of toiles in buildings;
  - maximum horizontal travelling distance;
  - minimum dimensions of cubicles;
  - maintaining privacy;
  - approach and transfer for the use of toilets;
  - door requirements;
  - the placement of sanitary equipment (toilet and toilet seat, wash hand basin, hand drier, grabrails, etc;
  - manoeuvring space and
  - vertical travel to an accessible toilet is limited to one storey.

- Toilets in Separate Sex Washrooms stipulates:
  - the provision of facilities for ambulant disabled people;
  - enlarged cubicles to for users requiring extra space, baby changing facilities and manoeuvring space;
  - the provision of equipment and grab bars and
  - minimum cubicle dimensions;

- Wheelchair-accessible Changing and Showering Facilities are not the subject of widespread use in the commercial property sector as toilets and washrooms. These are typically evident in the leisure sector with facilities located in hotels, sports centres and gyms. Increasingly, office accommodation has the provision for showers and changing facilities allowing abled-bodied staff or employees to cycle to work or undertake fitness activities. Accordingly, where such facilities are installed, these should also be fully accessible to those with disability. Approved document Part M of the building regulations stipulates the provisions for wheelchair-accessible changing and showering facilities to include:

- shower compartment layout;
- dimensions of shower cubicles;
- position and specification of equipment including emergency assistance call-cord and
- floor finishes.

- Wheelchair-accessible Bathrooms are more likely to be evident in the leisure and residential sectors of commercial property, accordingly the regulations stipulate:

- minimum dimensions;
- layout and transfer;
- floor finish;
- door openings and
- emergency assistance pull-cord.

British Standard 8300

As alluded to at the start of this chapter, British Standard 8300 is considered to be the *gold standard* with respect to the provision of an inclusive environment. The document is divided into two parts:

- BS8300-1: 2018 Design of an accessible built environment – Part 1 External Environment. Code of Practice.
- BS8300-2: 2018 Design of an accessible built environment – Part 2 Buildings. Code of Practice

While the Building Regulations are often associated with the provision of legal minimum requirements, British Standards are perceived as going above and beyond these. However, it is important to state that the adoption of British Standards and in particular those associated with providing an inclusive environment are not obligatory or enforceable in law. British Standard 8300 is published by the British Standards Institute (BSI) and according to the UK Government information on 'Standardisation',[17] the BSI is an independent body established in 1901 by Royal Charter. Fundamentally, there is a recognition that there is public interest in standardisation, and there is a need to apply this standardisation across a very wide range of products and processes. Within the UK government, the Department of Business, Energy and Industrial Strategy (BEIS) was formerly responsible for government general policy on standards. This government department has been replaced by two separate entities in 2023; the Department for Business and Trade (DBT) and the Department for Energy Security and Net Zero (DESNZ). Irrespective of the changes to government departments, there exists a *Memorandum of Understanding*[18] between the British Standards Institute (BSI) and the UK Government Department of Business, Energy and Industrial Strategy (BEIS). This identifies in *Article 2 The Public Policy Interest* as a driver for standards and within this are a number of key factors which can be contextualised with specifically British Standard 8300 for the provision of an inclusive environment:

- Standardisation is a factor supporting government policy on:
  - competitiveness;
  - innovation;
  - fair trading and
  - protection of consumer interests.
- Standardisation when used in conjunction with legislation can promote better regulation.
- There is a recognition that market forces and public interest are not always aligned with a need to address the challenge that innovated organisations may deviate away from standards to promote their own specification to exclude competition.
- A recognition that slow and out of date standards may be a hindrance and that standards can create barriers as well as removing them.

What appears obvious from the *Memorandum of Understanding* is that there is a commercial aspect to the positioning of standards and therefore more applicable to the production of goods over the delivery of service. Concerning disability and access to goods as well as services in the built environment, it can be concluded that adopting standards improves commercial competitiveness. This resonates with the findings of research into Public and Commercial Attitudes to Disability in the Built Environment[19] which identifies commercial benefit in providing an inclusive environment. From a user perspective, standardisation as identified in the memorandum protects consumer interests, and when applied in conjunction with legislation, it can promote better regulation. This is evident with the obligation placed on service providers with the Equality Act 2010 to make reasonable adjustment to facilitate access to goods and services for those with a physical or mental impairment. Furthermore, it is noted within the Building Regulations Approved Document Part M on numerous occasions, this defaults to the guidance provided by British Standard 8300.

Importantly, there is recognition that out-of-date standards can be a hindrance, and accordingly, there is evidence of regular periodic revision of BS8300. However, unlike standards affecting the manufacture of products, BS8300 provides guidance for inclusivity. In essence, the types or category of disability have remained largely unchanged for decades or even centuries, and while it is necessary to recognise that medical advances have identified a range of specific conditions, all disabilities default under one of the following categories:

- mobility;
- stamina and breathing;
- dexterity;
- mental health;
- memory;
- hearing;
- vision;

- learning;
- social and behavioural and
- other (including hidden disabilities).

Furthermore, when accessing the built environment for goods, services and employment the following access criteria also remain largely unchanged and are likely not to change radically in the future:

- access route
- parking;
- main entrance;
- horizontal circulation;
- vertical circulation and
- sanitary provision/facilities.

To position the role that British Standard BS8300 Parts 1 and 2 play in the provision of an inclusive environment above and beyond that of Approved Document Part M, it is necessary for further analyse these documents:

BS8300-1: 2018 Design of an accessible built environment – Part 1 External Environment. Code of Practice.

To position BS8300 Parts 1 and 2 in the process of contributing to inclusive design, it is necessary to re-affirm the point that this guidance is not a legal requirement. Furthermore, there is no doubt that the application of the document is intended for use with new build projects or significant redevelopment. This is evident in Section 4 of the document, which explicitly details integrating inclusive design principles into the development process. It is always likely to be more straightforward adopting the guidance for inclusive design with the 'blank canvas' of a new build or redevelopment. Therefore, adopting or upgrading the existing built environment for inclusive design is always likely to result in compromise.

Section 5 of BS8300-1 further details the optimising of an existing site such as the natural contours when planning a new build property to facilitate access but again this strategy is less likely with existing buildings. Certainly, with town or city centre commercial property, the notion of utilising the existing gradients to facilitate the access route is unlikely to be possible. BS8300-1 identifies the notion that access and wayfinding should adopt the *Principle of two senses* which in essence equates to designing for two of the following senses:

- audible;
- tactile and
- visual.

When seeking to implement retrospective upgrade for accessibility, there appears no discernible reason not to utilise a *two senses* approach. The technical detail of BS8300-1 is mostly in the following sections:

- Section 6 – Detailing Destination Arrival;
- Section 7 – Parking Provision;

- Section 8 – Horizontal Access and Approach Routes;
- Section 9 – Vertical Movement; and
- Section 10 – Way Finding through Public Space.

Section 6 of BS8300-1 concerns destination arrival and in particular the provision of a 'drop off' point as well as the requirements for bus stops, train platforms and bus/tram shelters. In the context of delivering access to commercial property, goods, services and employment, very few buildings have their own bespoke bus stop, train platform or tram shelter, other than transport hubs. Accordingly, it is necessary to consider that the destination arrival will concern mostly drop off points for taxis or private motor vehicles. With this in mind, Section 6 outlines the requirements for best practice, and it should be noted the tangible difference between Approved Document Part M of the building regulations and BS8300-1:

Approved Document Part M (Summary)	BS8300-1: 2018 (Summary)
• Clearly signposted setting down point. • Firm and level ground. • Close as practicable to the principal entrance. • Surface level with the carriageway.	• Designated setting down point on firm and level ground. • Close to accessible building entrance. • Clearly indicated. • Setting down point additional to designated accessible parking places and taxi drop point. • If feasible short-term waiting provision should be facilitated for drivers waiting & picking up. • Setting down point should have minimum dimensions of 9 m x 3.6 m. • Where feasible the setting down point should be covered and at a minimum height of 2.6 m. • Dropped kerbs should be provided. • The surface of access routes adjacent to the setting down point should allow for convenient transfer to and from a wheelchair. • Parking within a building should have a setting down point adjacent to the access route to the principal entrance.

As indicated above the prescriptive difference between the Building Regulations Approved Document Part M and BS8300-1: 2018 is vividly evident in the provision of a setting-down point. This philosophy is widely adopted throughout when a comparative analysis is undertaken between the two documents. This is why the British Standard is considered to be 'best practice' as it goes above and beyond that of the Building Regulations.

Section 7 of BS8300-1: 2018 concerns parking provision, a summary of the following key features in this section of the document:

- dedicated accessible parking spaces for employees;
- between 5% and 6% of overall car parking provision should be accessible (dependent on building use);
- between 4% and 5% of overall car parking provision should contain enlarged spaces (3.6 m x 6 m);

- on street parking design criteria;
- off street (car park) design criteria including dimensions, markings and signage;
- multi-storey car parking design criteria along with garaged spaces with one of the key requirements being a height clearance of 2.6 m to accommodate hoist facilities and
- ancillary parking requirements such as the design criteria for barriers and ticket machines.

Adopting a similar approach to the Building Regulations Approved Document Part M, the British Standard addresses horizontal access or approach routes in Section 8. The general design principles can be applied to town or city centres, as well as access routes within private commercial business site. These identify general requirements for blind or visually impaired people, such as being able to identify hazards with a cane as well as adopting contrasting colours between the access route and object. Section 8 emphasises the importance of providing continuous accessible routes between different buildings or from accessible parking spaces. The specific design criteria for horizontal access routes include requirements for the following:

- width of access routes including the need for passing places where this is too narrow or obstacles are present;
- gradients and cross fall of access routes;
- design criteria addressing the hazards attributable to the presence of street furniture;
- the placement of information and signage;
- pedestrian surfaces and the need to address

    - firm, slip resistant and smooth surfaces;
    - the avoidance of cobblestones, bare earth or gravel;
    - joints between paving and
    - appropriate use of tactile and deterrent paving.

- gates, barriers and restrictions

    - opening and closing mechanism should be accessible;
    - clear opening dimensions stated.

Section 9 of BS8300-1 details the requirements for vertical movement in the context of the approach to buildings, as well as external way finding and is divided into a number of detailed, prescriptive subsections including:

- Vertical movement (external steps, stairs, ramps, escalators and lifts)

    - weather protection and covering of stairs;
    - dimensions or treads, risers and overhangs;
    - maximum recommended numbers of risers;
    - the provision of steps additional to ramps;
    - avoidance of a single step;
    - stair width dimensions;
    - the provision for non-slip nosings;
    - landings (dimensions, gradient and cross falls);
    - lighting levels;

- ramped access;
- ramp location signage and directional signage to the principal entrance;
- ramp gradients (slope);
- ramp widths;
- landing dimensions;
- edge protection to ramps, surface materials and lighting;
- handrail principles for steps, stairs and ramps;
- handrail dimensions, fixings and material properties and
- explicit reference to BS8300-2 but general principles detailed.

BS8300-1: 2018 is wider ranging document than Building Regulations Approved Document Part M, and accordingly, it includes the provision of access to that go beyond buildings. This includes way finding through public spaces, and this is illustrated in Section 10 of the document, which details inclusive design to public meeting spaces, car parks, visitor attractions, temporary events, public art, outdoor cafes, public telecommunications, seating, water features landscaping and public toilets. In the context of accessibility to commercial goods, services and employment, this section of the British Standard has less relevance. The penultimate section on Part 1 of BS8300-1 concerns lighting, and the criteria for this appear in some for in most of the separate sections of the document. The lighting requirements for the provision of an inclusive environment cover the following sub-sections:

- general lighting principles which accommodate mostly the needs of blind or partially sighted people, the deaf and some neurological conditions;
- minimum and average lighting levels;
- avoiding glare or shadows;
- colour rendering;
- illumination for lip reading;
- ramps, steps and stairs;
- meeting and information points and
- wayfinding.

Prior to the information contained in the appendix is the final section of BS8300-1: 2018, and this concerns specific external locations such as:

- nature trails;
- beaches and piers;
- parks and gardens;
- fishing and angling;
- historic landscapes and monuments and
- Play areas.

These are all bespoke locations with often highly individual layouts as well as being predominantly external. Despite this, where there are buildings included within these locations, much of the same principles should be applied concerning access. Chapter 8 of this book will specifically address the provision of accessibility to the historic built environment. However, there is a flavour of the compromise that is required when inclusivity legislation meets building preservation. There is a measured discussion in this section of the standard which recognises the benefit of an inclusive approach to the historic built environment and the benefits an access for all approach can deliver. However, this is tempered by the need to contextualise the historical value in the delivery of a balanced approach to making alterations or reasonable adjustment.

BS8300-2: 2018 Design of an Accessible Built Environment – Part 2 Buildings. Code of Practice

Part 2 of BS8300 concerns primarily 'Buildings' and one of the key observations of the difference between the Building Regulations Approved Document Part M and BS8300 as a whole is the size of the documents. Approved document Part M comprises 74 pages stipulating what is perceived to be the legal minimum, in contrast the combined Parts 1 and 2 of BS8300 is 332 pages. This reflects primarily the more detailed approach and prescriptive nature of the British Standard. The definition of 'Buildings' is located in Section 20 of the document which is the last section before the appendix. In essence the document categories and defines buildings to include those within the following categories:

- transport related buildings;
- industrial buildings;
- administrative and commercial buildings;
- health and welfare buildings;
- sports-related buildings;
- religious buildings and crematoria;
- educational, cultural and scientific buildings;
- historic buildings;
- travel accommodation and venues and
- shops, supermarkets and shopping malls.

These categories of buildings listed within BS8300-2: 2018 are similar to the categories of buildings detailed in Chapter 5 of this book. The definitions of the commercial properties in BS8300-2: 2018 are quite general, whereas Chapter 5 looks deeper into the sector specific characteristics of the building types, and in particular, the issues to be considered when retrospectively making 'reasonable adjustment' to facilitate access.

In line with the framework of the standard laid out in BS8300-1: 2018, Part 2 of this British Standard sets out the rationale and detail for providing an inclusive environment. The early sections of the document are used to introduce some of the general inclusive requirements, the more technical sections are arranged as follows:

- Section 6 – Destination Arrival and Parking Provision;
- Section 6 – Access Routes to and Within Buildings;
- Section 8 – Building Entrance;
- Section 9 – Horizontal Movement;
- Section 10 – Vertical Movement;
- Section 11 – Surface Finishes;
- Section 12 – Signage and Information;
- Section 13 – Audible Communication Systems;

- Section 14 – Lighting;
- Section 15 – Facilities in Buildings;
- Section 16 – Counters and Reception Desks;
- Section 17 – Audience and Spectator Facilities;
- Section 18 – Sanitary Accommodation; and
- Section 19 – Individual Rooms.

Sections 6 and 7 concern arriving at a destination and the parking provision as well as the access to and within buildings. This part of the British Standard refers the reader explicitly to the relevant information contained within BS8300-1: 2018.

Section 8 of B8300-2: 2018 addresses the building entrance recognising that this can often present a barrier to access. Accordingly, the design criteria stipulated in this section covers the following to facilitate an inclusive environment:

- visual clarity of any entrance door to include signage, lighting and contrasted against the adjacent façade;
- weather protection such as recessed entrances or protective canopies;
- door threshold;
- entrance doors:

  - self-closing swing doors;
  - power operated doors;
  - revolving doors;

- entrance lobbies and airlocks (dimensions, glazing and projections);
- door width dimensions (at narrowest point including door furniture);
- door side clearance;
- manual sliding doors;
- visual contrast of doors and walls;
- vision panels and glazed door;
- door fittings and furniture;
- controlled door closing devices (force required to manually open doors);
- hinges, locks and latches;
- panic bars, buttons and emergency exit devices;
- access control systems (including door entry systems, entry phones, turnstiles and security pass gates);
- entrance and reception areas including prescriptive guidance on:

  - clear signage (particularly to multi-tenanted buildings) and directional signage;
  - wayfinding provision where there is no reception point;
  - floor finishes including the flooring system to remove debris and water from shoes and wheels and
  - floor finish requirements to meet with Section 11 of BS8300-2: 2018.

- reception points:

  - clearly identifiable;
  - facilitate both deaf or hard of hearing as well as blind and visually impaired;
  - reception desks visually contrasting and not in front of pattered background;

- provision of seating;
- approach to reception from entrance point should be free from obstruction, firm and slip resistant;
- appropriate signage and universal pictograms for staircases, lifts and emergency escape routes.

- interview rooms (minimum dimensions, turning space and clear door width).

Adopting a similar logical approach to that of Building Regulations Approved Document Part M, the criteria for inclusive design in BS8300-2 identifies horizontal movement in Section 9 following on from the requirements for the entrance noted in Section 8. This in essence details:

- minimum internal corridor widths which in principle should permit wheelchair users to the ability to turn 180°;
- minimum dimensions to allow for two wheelchair users to pass each other;
- avoiding projections into corridors and passageways including the placement of protective barriers if necessary;
- floor finishes to corridors;
- lighting levels in corridors;
- doors:

  - opening into corridors;
  - placed within corridors;
  - automatic door closing mechanisms;
  - fire doors with controlled door closing devices;
  - non-fire doors with controlled closing (for security); and
  - door fittings.

- fire escape routes.

Section 10 of BS8300-2: 2018 concerns vertical movement and makes explicit recommendations for steps, stairs, ramps, handrails and lifting devices. There are a number of prescriptive recommendations addressing vertical movement or circulation, and these essentially divided into the following:

- steps and stairs;

  - dimensional recommendations for risers and goings (treads).
  - limitation on number of risers per staircase or flight;
  - uniformity in riser height for individual staircases or steps;
  - avoiding single steps (to be replaced with ramp);
  - identification and slip resistance of nosings;
  - placement of landings;
  - lighting levels to ensure it is possible to identify the difference between each step and riser;
  - lighting provision to avoid glare;
  - surface materials to staircases;
  - the placement of refuges within fire compartments.

- ramps and slopes:

  - recommended gradients;
  - landings (placement and dimensions);
  - ramp widths and edge protection;
  - surface material (non-slip and colour contrasting with landings and edge protection);
  - lighting;
  - temporary and portable ramps (noted portable ramps should only be used in exceptional circumstances and this might be a valid compromise with the historic built environment).

- handrails to ramps and steps:

  - principles of design: Easy, comfortable grip with no sharp edges, continuously graspable, finished to provide a visual contrast with surroundings, extending beyond the first/last step, terminated to avoid catching clothes and strong enough to support users;
  - specific recommended dimensions and spacing;
  - hazard protection beneath stairs and ramps.

- lifting appliances

  - key statement that lifts provide the only method for vertical circulation for a number if disabled building users and at least one conventional lift should be installed;
  - Section 10 covers a wide range of lifting appliances;
  - accessing lift appliances and the location should be clearly indicated from the building entrance;
  - sufficient manoeuvring space should be provided to the lift access point with sufficient lighting;
  - conformity of the lift should be with both separate British Standards and European Norm;
  - minimum car sizes are defined;
  - design criteria for the placement of lift call buttons to the lift lobby, controls and door width with emphasis on visual contrast of the buttons, mounting plate and floor number marking on the wall opposite the lift door;
  - advice for slip resistant floor finishes and a frictionless transition between the lift car floor and the landing;
  - colour contrast of floor finish with lift car walls and avoiding where possible glass elevators.
  - noted is a requirement under BS8300-2 Section 10 to foresee audible and visual indicators for destination arrival as well as accessible communication system including variable audio microphone and the presence of an induction loop;
  - specific design criteria including appropriate independent power supply for accessible lifts purposed to function for emergency evacuation;
  - lifting platforms (both open and enclosed) are detailed including minimum dimensions with similar requirement for colour contrast and stipulated door width;
  - a recognition that lifting platforms usually travel one storey or 3 m;

- further design recommendations for wheelchair stair lifts which are considered to be the least appropriate installation behind conventional lifts and platform lifts;
- stairlifts pose a risk concerning blockage of emergency staircase and there is note specific design criteria on how to maintain clear access;
- differing from Approved Document Part M, there is design criteria for escalators and walkways although a recognition that are hazardous for wheelchair users and those with disability.

Section 11 of BS8300-2: 2018 is dedicated to surface finishes and relates to many aspects of the accessibility features already discussed. The design and choice of materials as well as the surface finish may have a significant impact on accessibility for the deaf and those with hearing loss, sight loss and potentially everyone using mobility aids. Surface finishes within the built environment broadly equate to those placed on floors, walls and ceilings they are important because they contribute to the following:

- colour and contrast (including patterns);
- reflection – light;
- reflection – sound and
- texture.

Colour and contrast are of benefit to many disabilities. They are typically used for way finding as well as signage. It may be stated that good colour contrast is also good practice with respect to the lettering or pictograms on signage for all built environment users. However, less attention is paid to the requirements for visual contrast between floor finishes and wall, as well as wall finishes and ceilings; this is often the case with internal fit-out schemes. Architectural design and the specification of finishes are invariably driven by aesthetics; this can have a negative impact on accessibility. The British Standard stipulates a numerical value of contrast, which is measurable and good design can maintain aesthetic while also providing appropriate contrast. Generally patterned surfaces are not good inclusive design for floor finishes and can create the impression of steps or undulations to those with sight loss.

One key requirement of finishes is that they do not create glare or reflection, and this can be a combination of the reflective characteristics of the material with the presence of both natural or artificial lighting. This can prove challenging for designers, particularly with the use of glass, metals and ceramics to internal fit out. Reflection is also a factor with acoustics, and the finishes to surfaces should be designed to limit reverberation. It is recognised that acoustics, and the passage of sound is also dependent on the enclosure of spaces such as auditoria, railway stations or building lobbies. The science and design attributable to the reflection of sound is complex but can have a significant benefit for the deaf or those with hearing loss.

The texture of surfaces is largely implemented in the terms of accessibility for way finding, limiting risk or providing warning. This is apparent with items such as 'blister paving', which is placed at the crossing points of roads or textured surfaces applied to the nose of stair treads as a deliberate act to engender friction to prevent slips and falls. Despite this, changes in the texture of a surface can cause a problem, and this is evident with the placement of a floor mat in an entrance hall where the surface either side is relatively frictionless. Texture is an important concept in the terms of creating an inclusive environment but there is a nuance required regarding its application.

Section 11 essentially addresses finishes with respect to their use and characteristics in the following:

- Floor surfaces – Shine, glare and slip resistant with recommendations to avoid floor joints or exposed edges causing a trip hazard. While carpet and carpet tiles are acceptable, deep pile carpet should be avoided.
- Wall surface – Should avoid bold patterns, particularly behind reception desks where plain backgrounds cause less distraction for lip reading. Essentially plain finishes are recommended with surface-mounted sockets, outlets or other functional elements contrasted in colour (as well as texture) against the background.
- Glazed walls and Screens – The obvious risk is that glazing has the appearance of unimpeded access to those with sight loss and poses a health and safety risk. Manifestation (surface etching or markings) should be placed on the surface of the glass and this should contrast against the background behind the glazing panel. Glass with low light reflectance should be incorporated.

Section 12 of BS8300-2: 2018 addresses signs and information; there is a recommendation that this should include visual, audible and tactile ways to communicate the necessary information associated with the purpose and layout of spaces. There is a need to consider the location of signage as well as the positioning of this including height. As discussed on numerous occasions, colour contrast is important, and this is in the context of any lettering or symbol against its background. Full inclusivity embodies the integration of audible and visual communication.

It is noted that directional signage is an important requirement, and this should be placed at the junctions of access routes or fundamental locations such as a building reception or the sanitary rooms. Tactile signage should be positioned at a suitable height for all, including wheelchair users; these should be positioned to lift lobbies, staircases and reception areas.

The information contained in Section 12 is highly prescriptive and includes the relevant dimensions of lettering or numbers as well the recommended positioning of these to walls and surfaces. This section is completed with reference to the use of audible information complimenting visual and tactile signage. This is the lead into the following section which specifically addresses the implementation of audible communication systems.

Section 13 addresses the necessary systems to enhance audible communications for the deaf or those with hearing loss. It identifies a range of applications and advises on the principles of these. In summary, audible communications include public address systems, which should include assistive listening facilities. The key issues concerning the clarity of audible communications are the avoidance of amplifying background noise and the typical problems that occur with reverberation to enclosed spaces. One way to reduce this issue and enhance audible communication is with the presence of enhanced listening systems or hearing loops. The standard addresses in particular the installation of hearing loops which allow users to switch their hearing aids to the 'T' setting to improve clarity of communication. These work with a microphone, amplifier and wire loop which creates a magnetic field. This converted with 'T' setting on a hearing aid to receive the sound signals, and while it is considered a relative low tech but effective tool, it does not differentiate between different signals. Therefore, if there are a number of induction loops in a confined location, interference is possible.

Section 13 also details infrared systems and radio or Wi-Fi communication, these are considered to be more sophisticated modes of audible communication. In the case of infrared, these involve individual user headsets, which can be looped to hearing aids and typically have more than one channel. The application of these may be typically in cinemas where one channel is used for enhancing audio and the other for providing audio description for those with sight loss. Radio and Wi-Fi communication essentially require a user to have a receiver or device to pick up audio, and this technology is widely used on museums or visitor attractions where commentary can be strategically picked up by users during their visitor experience. Increasingly, Wi-Fi connectivity through App technology on personal devices is another way to deliver one to one audible communication.

An important aspect of Section 13 is the provision of alert or alarm systems, and the standard specifies the detail of these. One clear recommendation is for visual alarms, which are typically used to indicate activation of a fire alarm, but there is also a need for an alert system to be installed within sanitary rooms or cubicles. It is also necessary for the provision alert and communication to be evident in refuge locations. These are typically within separate fire compartments, such as staircases which are equipped with fire doors and appropriate smoke extraction. Such facilities are to provide emergency refuge where it is not possible to for accessible escape provision. While refuges may be deemed a reasonable adjustment in the context of the existing built environment, there should be no conceivable reason why there are no accessible escape routes to new buildings. The lives of occupants should not be compromised in the event of an emergency because they are disabled.

Section 14 of BS8300-2: 2018 addresses one of the most evident requirements for inclusivity and this is lighting. It is clear that there are so many facets to creating an inclusive environment, and it has already been discussed colour, contrast, and texture in the context of specific individual accessible features. However, without basic lighting or illumination, user access of the built environment as a whole would not be possible and accordingly this is a critical consideration within inclusive design.

The standard addresses the key principles of lighting design, and while Section 14 is not prescriptive in nature, the document as a whole contains lots of references to the principles as well as the criteria. This is evident with all aspects of access from parking provision through to vertical circulation. There are examples of the necessary levels of lighting which are recorded in Lux but the key principles for illumination concern the following:

- good lighting is critical for the partially sighted and those with neurological or sensory processing difficulties.
- lighting should consider:
  - the reflectance on floor, wall and ceiling finishes;
  - appropriate levels of illumination for activity such a lip reading;
  - avoidance of glare, pools of light and shadows;
  - avoidance of uplighters, floor or low-level lighting;
  - illuminance to maintain colour rendering with a recognition that strong colours can cause sensory overload;

It is the application of these principles in the context of the different accessible features that contribute to and influence inclusive design.

Sections 15 to 19 may be considered the most prescriptive of BS8300-2: 2018 in the sense that these contain a high quantity of annotated diagrams detailing key dimensions. These sections also detail accessible issues that are most likely to affect the users experience, including the following:

- Section 15 – Facilities in Buildings:

  - seating and waiting areas;
  - storage facilities;
  - ATMs and cash-operated devices;
  - touch screens;
  - windows and their operation;
  - public telephones;
  - outlets, sockets and switches and
  - assistance dog facilities.

- Section 16 – Counters and Reception Desks:

  - positioning and access;
  - desk dimensions, space, surface and acoustics;
  - control barriers for queuing and
  - space for secure transactions.

- Section 17 – Audience and Spectator Facilities:

  - seating;
  - wheelchair spaces;
  - access to audience seating;
  - raked floors;
  - ancillary equipment and
  - lecture and conference facilities.

- Section 18 – Sanitary Accommodation:

  - shower rooms and bathrooms;
  - changing and shower areas;
  - accessible baby changing facilities;
  - toilet accommodation and
  - changing places toilets.

- Section 19 – Individual Rooms:

  - kitchen areas;
  - accessible bedrooms and
  - quiet spaces.

In the context of providing access to commercial property, goods, services and employment commercial properties will encompass the requirements detailed in Sections 15 to 19 in accordance with the destination use of the space or operator. For example, it is rare for retail facilities to have a reception desk, but they will invariably need an accessible counter for secure transactions. The requirements stipulated for accessible bedrooms also are only relative to a specific use class. While the provision of accessible toilets offers a clear example of attempted inclusivity, the guidance simply states that those with disability should be able to find and use suitable toilet accommodation no less easily than non-disabled people. It does not recommend widespread implementation of accessible toilets if these are not a normal requirement for the commercial service provider. Accordingly, there is no standard requirement for retail units to provide sanitary facilities and therefore no requirement to install accessible facilities.

Implementing Legislation and Guidance in the Built Environment

While there exists both legislation determining minimal legal compliance and guidance addressing best practice for inclusivity, there exists a requirement to contextualise so much of this. It is critical not to deviate from the base principle in The Equality Act 2010 that concerning access to goods, services and employment, it is not possible to discriminate on the basis of disability. Therefore, everything that is available to those without a disability should be the same for those with a physical or mental impairment.

To begin to understand the challenges and opportunities presented by providing an inclusive environment, it is necessary to define commercial property as well as analyse the different characteristics that exist within the different commercial property sectors. Chapter 5 of this book looks at this in detail and recognises that there is significant diversity in the building types, added to this is the complexity of retrofitting for accessibility. This is further muddled by the use of the term 'reasonable adjustment'. Therefore, prior to recommending methods to achieving accessibility, it is always necessary to examine the existing building and its features with reference to the requirements. This is the process of audit and will be discussed fully in Chapter 6 of this book; however, in essence, an audit is the measure of the status of current accessibility against an appropriate benchmark. Chapter 7 discusses public and commercial attitudes to disability in the built environment; there is a notion that with respect to compliance and the need to deliver 'reasonable adjustment', the benchmark should deliver legal compliance. This is not uncommon in the built environment, and even Approved Document Part M of the building regulations concedes that a building may be in compliance but still not accessible. While those investing in commercial property or delivering goods, services and providing employment to those with disability appear conditioned to accepting minimal compliance. Contrary to this and the 'race to the bottom', research has established that there is commercial value in creating a truly inclusive environment. The fundamental message for all those providing access to goods, services and employment is to choose carrot over stick and opportunity over obligation. Accordingly, and recognising that there will always be some compromise with retrofitting access to existing buildings, this should strive, where possible, to adopt best practice.

Notes

1 Fiftieth anniversary of the Chronically Sick and Disabled Persons Act 1970 – House of Lords Library (parliament.uk)
2 Chronically Sick and Disabled Persons Act 1970 (legislation.gov.uk)

3   National Assistance Act 1948 (legislation.gov.uk)
4   DDA 25 years on: 'Phenomenal activism … but deeply flawed legislation' – Disability News Service
5   The public's perception of disabled people needs to change — we're not just Paralympians or scroungers | The Independent | The Independent
6   Fleck, J. (2019). Are You an Inclusive Designer? RIBA Publishing.
7   Equality Act 2010 (legislation.gov.uk)
8   Definition of disability under the Equality Act 2010 – GOV.UK (www.gov.uk)
9   Equality Act 2010 Guidance (publishing.service.gov.uk)
10 employercode.pdf (equalityhumanrights.com)
11 Firstgroup Plc v Paulley | Disability Rights UK
12 FirstGroup Plc (Respondent) v Paulley (Appellant) (supremecourt.uk)
13 DDA 25 years on: 'Phenomenal activism … but deeply flawed legislation' – Disability News Service
14 employercode.pdf (equalityhumanrights.com)
15 Approved_Document_M_vol_2.pdf (publishing.service.gov.uk)
16 Approved_Document_M_vol_2.pdf (publishing.service.gov.uk)
17 Standardisation - GOV.UK (www.gov.uk)
18 Memorandum of Understanding Between BSI and BEIS (publishing.service.gov.uk)
19 Public-and-Commercial-Attitudes-to-Disability-in-the-BE.pdf (ucem.ac.uk)

References

Blunkett, D. (2015). *The public's perception of disabled people needs to change — we're not just Paralympians or scroungers*. The Independent [online]. Available at: https://www.independent.co.uk/voices/comment/the-publics-perception-of-disabled-people-needs-to-change-were-not-just-paralympians-or-scroungers-10104704.html [Accessed 14 January. 2024].

Disability News Service. (2020). *DDA 25 years on: 'Phenomenal activism … but deeply flawed legislation'* [online]. Available at: https://www.disabilitynewsservice.com/dda-25-years-on-phenomenal-activism-but-deeply-flawed-legislation/ [Accessed 14 January 2024].

Disability Rights UK. (n.d). *Firstgroup Plc v Paulley* [online]. Available at: https://www.disabilityrightsuk.org/firstgroup-plc-v-paulley#:~:text=Supreme%20Court%20decision&text=The%20Supreme%20Court%20unanimously%20allows,any%20further%20steps%20was%20unjustified [Accessed 14 January.2024].

Equality and Human Rights Commission. (2011). *Equality Act 2010 Code of Practice – Employment Statutory Code of Practice* [online]. Available at: https://www.equalityhumanrights.com/sites/default/files/employercode.pdf [Accessed 14 January. 2024].

Fleck, J. (2019). *Are You An Inclusive Designer?*. RIBA Publishing.

Gov.UK. (n.d). *Chronically Sick and Disabled Persons Act 1970*. [online]. Available at: https://www.legislation.gov.uk/ukpga/1970/44/contents [Accessed 14 January 2024].

Gov.UK. (n.d). *Definition of Disability under the Equality Act 2010* [online]. Available at: https://www.gov.uk/definition-of-disability-under-equality-act-2010 [Accessed 14 January. 2024].

Gov.UK. (n.d). *National Assistance Act 1948* [online]. Available at: https://www.legislation.gov.uk/ukpga/Geo6/11-12/29/enacted [Accessed 14 January 2024].

Gov.UK. (n.d). *The Equality Act 2010* [online]. Available at: https://www.legislation.gov.uk/ukpga/2010/15/contents [Accessed 14 January 2024].

Gov.UK. (n.d). *Memorandum of understanding between bsi and beis* [online]. Available at: https://assets.publishing.service.gov.uk/government/uploads/system/uploads/attachment_data/file/910301/memorandum-of-understanding-between-bsi-and-beis.pdf [Accessed 14 January 2024].

Gov.UK. (2020). *Guidance Standardisation* [online]. Available at: https://www.gov.uk/guidance/standardisation [Accessed 14 January 2024].

HM Government. (2020). *The Building Regulations 2010, Approved Document Part M, Access to and use of buildings, Volume 2 -Buildings other than dwellings* [online]. Available at: https://assets.publishing.service.gov.uk/government/uploads/system/uploads/attachment_data/file/990362/Approved_Document_M_vol_2.pdf [Accessed 14 January 2024].

Office for Disability Issues HM Government. (2011). [online]. Available at: https://assets. publishing.service.gov.uk/government/uploads/system/uploads/attachment_data/file/570382/ Equality_Act_2010-disability_definition.pdf. [Accessed 14 January 2024].

Tagg. (2020). *Public and Commercial Attitudes to Disability in the Built Environment* [online] Available at: https://www.ucem.ac.uk/wp-content/uploads/2020/10/Public-and-Commercial-Attitudes-to-Disability-in-the-BE.pdf [Accessed 14 January 2024].

The Supreme Court. (2017). *JUDGMENT FirstGroup Plc (Respondent) v Paulley (Appellant)* [online]. Available at: https://www.supremecourt.uk/cases/docs/uksc-2015-0025-judgment.pdf [Accessed 14 January 2024].

UK Parliament House of Lords Library. (2020). *Fiftieth anniversary of Chronically Sick and Disabled Persons Act 1970* [online] Available at: https://lordslibrary.parliament.uk/fiftieth-anniversary-of-the-chronically-sick-and-disabled-persons-act-1970/ [Accessed 14 January 2024].

Chapter 5

Commercial Property Sectors, Service Providers and Access to Buildings for Employment

Introduction

Broadly speaking, property is either classed as 'residential' or 'commercial', and in essence, the provision of access to goods or services as well as employment is associated with commercial properties. To begin to understand the different commercial buildings, operations or sectors, it is necessary to establish the definition of commercial property.

What sets commercial and residential property apart is that commercial property is essentially land or buildings used in an act of commerce. This could be interpreted as a shop where trading occurs, but more intrinsically, it covers a wide range of activities that are encompassed under the title of 'businesses'. Commercial property does not have to be profitable, and this might typically occur with public sector assets or even professional football clubs that can generate millions of pounds of income yet run at a loss. For the sake of providing an inclusive environment, commercial property is considered to be buildings where goods or services are provided with or without some kind of financial remuneration. When considering commercial property, it is, however, also necessary take into consideration residential property which is operated for business use.

A large percentage of residential property is owner occupied, and clearly, this is not commercial; however, it is appropriate to consider the private rented sector as commercial properties. Buy-to-let properties are unlikely to be considered as commercial properties as they are almost stand-alone investments. However, should buy-to-let owners acquire portfolios of residential properties, particularly if these are incorporated into the assets of a registered company, then these could then be classed as commercial properties. A good example of commercial properties in the residential sector are the housing stocks owned and maintained by local authorities or housing associations. This is when it is necessary to consider the application of many commercial regulations, including the provision of access. Furthermore, when looking at the UK Building Regulations Approved Documents, there is also a defined separation to allow for 'buildings other than dwelling houses'. These volumes of the regulations are intrinsically concerned with commercial buildings.

When considering commercial property, it is recognised that there are a wide variety of buildings but these can be put into the principal sectors:

- Residential.
- Retail.
- Offices.

DOI: 10.1201/9781003239901-5

Figure 5.1 Commercial property sectors and sub-sectors.

- Industrial.
- Leisure.
- Public sector / infrastructure.

Within each sector, a sub-division can be made to further illustrate the types of properties which make these (Fig 5.1).

Commercial property is either owned as an investment, occupied by a tenant or owner occupied. Before looking in more detail at the individual commercial sectors and their characteristics, it is worth considering the reasons why organisations invest in real estate. Banks, pension funds and other large investors will have a variety of investment vehicles in which they will seek to grow capital. While stocks and shares may be considered to have potential high returns on investment, they may also be considered high risk. Property, alternatively, may be perceived as being relative low risk, and although the returns may also be relatively low there, the property or building itself is a tangible asset.

Owners, Investors and Service Providers

For property investment ,there needs to be an initial outlay of capital to purchase the property, but ultimately, there needs to be an end user to occupy this and commit to a long-term lease, thus providing an income stream for the investor. As with most businesses, there will be a requirement for the investor to assess the costs or overheads required to buy, run, and maintain the property. This is then offset against the return or income from the tenant or end user to derive ideally a profit or sometimes a loss.

The occupiers of commercial properties undertake some kind of business activity, and this is including a wide spectrum of activities, but, in most cases, those occupying commercial properties are employers or service providers. Ultimately the accessibility of commercial property is governed by legislation aimed at the provision of goods and services. However, while the access to goods and services is fundamental in an inclusive environment, the other equally important facet of access concerns places of employment. Accordingly, legislation is also in place to ensure that the workplace is accessible to those with a physical or mental impairment, and this includes the need to make adaptive changes accordingly. Therefore, when considering the inclusive access to commercial property, this needs to address all real estate, which houses employees as well as selling goods or delivering services. The concept of providing physical access to commercial property should apply consistently for those providing goods and services, as well as to those employed on the delivery side. It is agreed that this would appear to cover a significant quantity as well as variety of buildings; however, the principles of providing access are relatively straight forward and can be applied in a uniform manner. To establish how to assess the built environment and deliver appropriate advice on access, it is necessary to analyse the different property sectors and their individual sub categories of building use (Fig 5.2).

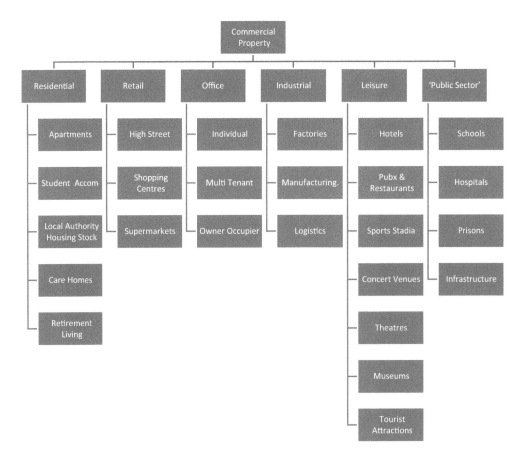

Figure 5.2 Sub-categories of commercial property.

Office

Aside from where we live, where we work is probably the place where the second-most amount of time is spent. While it is acknowledged that there are other important places of work concerned with manufacturing, infrastructure, logistics, retail, etc., offices are probably the sector which 'houses' most employees. In the terms of providing access to office buildings, this will be necessary in the context of those working or employed by companies residing in offices, as well as those visiting these organisations. All goods and service providers are obliged to make the necessary reasonable adjustments to ensure access is permitted. There appears some uncertainty over who is responsible for the implementation of the necessary works to install, manage and maintain access. This will often depend on the status of the employer or service provider operating out of the premises.

If this is an owner occupier, then they have sole responsibility for access to their goods and services; accordingly, they are responsible for undertaking the necessary reasonable adjustment to facilitate it. In the context of the physical requirement to provide building access, this is fundamentally separated into the following key components, which should also be considered for all categories of commercial building:

- Access Route (to the main entrance).
- Parking Provision.
- Main Entrance.
- Horizontal Circulation.
- Vertical Circulation.
- Facilities and Toilets.

Where the service provider or employer is a tenant, it is usual for them to occupy a defined space known as their demise. In essence, this is the limit or boundary of their space within a building that is defined in their lease agreement, typically with the property owner or the head lease. It is typical for a tenant's demise to exclude the main entrance to the building and the central core, which may be used to house the lifts and / or toilets as well as emergency escape staircases. These areas are known as the 'common parts', and while the tenant is responsible for accessibility within their demise, it is the building owner or landlord who is responsible for accessibility to the common parts. In some instances, the building may be let to a single tenant or multi-tenanted, but in both cases, it is the landlord who is responsible for the provision of access from the street, any parking provision, as well as the main entrance. These points of access are relatively straightforward to address, although the compliance of access to the main entrance could be potentially complicated and costly. Horizontal circulation is determined by the layout of the building reception and the route to the tenant's demise which is also evident on other floors if the building has more than one storey. Horizontal circulation within the tenant demise is charged at the responsibility of the tenant, but the biggest access obstacle and potentially significant cost is associated with vertical circulation if it is deemed necessary to install a lift. Toilets and facilities can be located within the common parts and sometimes within the tenant demise.

Primarily investors are seeking investment properties that will generate a yield (return) on their investment. The yield and commercialisation of the investment asset forms the principal advice of the investment/agency surveyors advising on the commercial due diligence. This will include the types and durations of lease contracts, price per square metre/foot as well as service charge details. All of this is used to establish the projected return on investment, which will ultimately determine the value or price of the asset. Naturally rental occupation and income vary significantly throughout the economic cycle; therefore, most investors will take this into consideration during the acquisition process and any subsequent negotiation.

The landlord/tenant relationship for commercial property can be fractious since despite both parties wanting their building or demise to be of the best possible quality and fit for purpose, they are conflicted about who should pay for it. This relationship is covered by the commercial lease between the two parties, as well as a number of legal acts. Outside of the acts covering the landlord and tenant relationship, there are also a number of imposed legislations to cover health and safety, fire and energy efficiency. Some of these are transferred onto the tenant for compliance, and others adopted in part by the tenant depending on how they operate or exploit the space.

While there is legislation in place obliging building owners to conform to Minimum Energy Efficiency Standards (MEES) to reduce the operational carbon footprint of a building, there is no explicit requirement for owners to provide an inclusive environment. However, it is up to the service provider to make the necessary reasonable adjustment to facilitate access to goods and services under the Equality Act 2010. This works typically where the tenant or exploiter is providing goods and services to the building as a whole if they are an owner occupier. However, when a tenant occupies a demise equating to part of the building or an office floor in a multi-let building, it is the landlord (as service provider to the tenant) who is now obliged to make reasonable adjustments to facilitate access (Fig 5.3).

Works to the access route, main entrance or common parts of an office building to facilitate access are typically funded by the landlord or building owner. Money invested

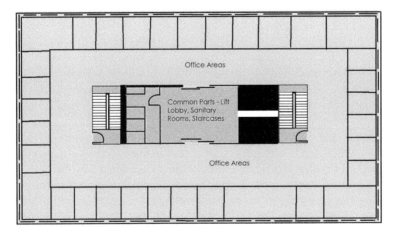

Figure 5.3 Illustrated in pink are the common parts of a multi-let office building, while the tenant (blue) is obliged to make reasonable adjustment to facilitate access to goods and services in their demise, the building owner/landlord is obliged to make reasonable adjustment to facilitate access from the common parts to the office floors.

in such works potentially affects their cash reserves and in part might go some way to influencing the approach to undertake the minimum legal standards. Alternatively, there is a notion amongst those investing and advising on commercial property to meet with the legal requirements, nothing more or nothing less. This is evident with the installation of fire doors and where the legal requirement is for say a 30-minute fire door, there is no commercial incentive to install a 60-minute alternative. This legal compliance makes sense although regrettably the same philosophy is adopted to the implementation of disabled access when going above the legal minimum does make a tangible difference.

Office Construction and Characteristics

Office buildings are mostly purpose built for this function and certainly post WW2, these are likely to resemble the office buildings that we know today in the sense that these are typically greater than three or four stories in height. Other office accommodation may form part of more mixed-use development, which may include retail or other functions on the ground floor with offices to the upper levels. Such office accommodation is typical to many historic towns or city centres. These properties may have been constructed pre-1920, and although used as offices, they are likely to have undergone renovation or change of use from their original function. Later in this chapter, it will be discussed in more detail the principal differences between pre- and post-1920s properties in the context of providing an inclusive environment.

'Modern' Office Buildings

When analysing modern office buildings, it should be noted that the concept of owner-occupier properties is something that is more prevalent with older buildings. In many cases, these buildings originate from bespoke businesses where head office buildings were constructed as a sign of importance or representative of sector dominance. Typically, these were constructed for the banking sector, pharmaceutical, engineering or oil industry. Alternatively, owner-occupier office buildings were constructed by central or local government at a time when there were sufficient amounts of funding available to do this. There was a sense of pride in the construction of council or civic buildings, which is less apparent towards the end of the 20th century.

In most cases, owner-occupied buildings are constructed to a high specification. It is supposed that this is due to the significance of the building to the owner's profile or to make an architectural statement portraying success, wealth, or the innovation of the company within its sector. Irrespective of this, there was so little attempt to consider inclusive design in the construction of offices until the 1970s. Furthermore, there was such a limited concept of disability and an absence in recognition of the diversity of this that the proposed access provision was more than limited. It was not until the advent of the Disability Discrimination Act in 1995 that it could be argued that the mainstream design or renovation of buildings adopted a joined-up approach to inclusivity. As a consequence, much of the purpose-built office buildings from the post-WW2 period through to the late 1990s is not compliant with current requirements for accessibility. However, due to material lifecycles and the cyclical nature of property investment, many of these buildings will have been the subject of demolition, redevelopment or renovation. Accordingly, it is anticipated that a more realistic attempt has been made to adopt

Figures 5.4–5.6 **Purpose built modern office buildings constructed and/or renovated post 2010.**

accessibility, even if this is to conform with the minimum legal requirements established in the Building Regulations, Approved Document Part M. It can be argued that stripping a building back to its frame and core should allow for the relevant compliance of basic accessible features associated with the main entrance as well as both horizontal and vertical circulation. However, the installation of a compliant passenger lift to accommodate vertical circulation is only possible if the existing shaft can facilitate it. Accordingly, any need to significantly alter the shaft dimensions can prove to be complex as well as costly.

Bespoke post-WW2 office buildings are usually located in the business districts of town or city centres and usually have good transport links although the principal access routes determined by the adjacent pavement or road networks. Some of these buildings have their own basement parking, and any assessment concerning accessibility should consider the numbers or accessible parking spaces as well as their dimensions. Clearly, there is a requirement for vertical circulation between any basement parking and the upper floors to be provided by a lift. As previously indicated, the dimensions of the lift are limited to that of the shaft, any requirement to alter this needs careful consideration.

Pre-1920s Mixed-Use (Offices)

Some office accommodation may form part of more varied mixed-use buildings, and while these may be modern and purpose-built developments, others are more localised refurbishments. This typically happens in town or city centres where older pre-1920s low-rise properties have a ground floor devoted to retail use and the upper floors as office accommodation. This is likely to be significantly more complicated to accommodate accessibility. When considering the access provision to these types of office buildings, it is important to have an understanding of the relevant construction technology as this is significantly different from that of post WW2 medium or high-rise offices. The different construction technology associated with pre- and post-1920s buildings is described in more detail later in this chapter. This construction technology knowledge of pre-1920s buildings used for offices is transferable between the different property sectors and particularly relevant with such buildings used for retail or hotels.

Essentially, pre-1920s buildings operated as office accommodation comprise similar construction characteristics as residential properties of the same age. Located mostly in town or city centres, these could occupy the frontage of former retail units or as detailed previously be situated at first-floor level over shops. The external facades may have been executed in brickwork or natural stone, but the concept of accessibility would not have been a factor at the time of construction. As a consequence, these largely Victorian or Edwardian buildings will have access typically directly off the street. Due to the nature of their age and location, they are highly unlikely to have parking facilities, and therefore, are accessed from the adjacent street. The access routes to these buildings are determined by their immediate surroundings, and it is the local authority who maintains these. Due to the town or city centre locations, pavement widths may vary; the materials used for pavement surfaces could be asphalt, concrete slabs, natural stone or cobbles. The type and quality of these materials may have a bearing on the accessibility of the principal building entrance.

The nature of pre-1920s properties used for offices is such that the entrance may be stepped off the street. This is a legacy of basic construction technology where internal floor levels are typically 150 mm above those of the external street level to prevent dampness and moisture penetration. Despite this, there are numerous examples or subsequent later alteration to entrances to provide ramped, step-free access (Fig 5.7).

Historic town or city centres may be considered areas or architectural significance and accordingly protected with *Conservation Area* status; furthermore, some pre-1920s buildings used for offices might be individually *listed*. These two legislative mechanisms may be used to protect or preserve the historical integrity of the buildings and might be perceived as a barrier in the provision of access. However, as discussed in Chapter 8 of this book, there are many excellent examples of retro fitting of entrances to ensure these are accessible.

Generally, the principal entrances to these buildings may be limited in dimension and access doors not wholly compliant with current requirements for disabled access. However, and subject to planning and or listed building consent, works are mostly possible to make reasonable adjustment to facilitate access. Having entered pre-1920s properties operated as commercial offices the horizontal circulation will be determined by the location of internal walls. While internal wall construction to modern offices is lightweight, flexible and non-loadbearing, the internal walls to many pre-1920s buildings operated as offices are structural. Therefore, these form part of the building stability and accordingly, the removal or alteration to these walls will require careful consideration as well as structural assessment. This can impact negatively on the widths of internal

Figures 5.7–5.9 Left (Fig 5.7) and top right (Fig 5.8): Pre1920s buildings with mixed use destination including office accommodation, typically with stepped access from street level. Bottom right (Fig 5.9): Post-WW2 bespoke office with its entrance raised above street level and with a retro fitted external access ramp.

corridors providing horizontal circulation and may test the notion of a 'reasonable adjustment' if it compromises structural stability.

Most pre-1920s buildings operated as offices do not have a lift to accommodate vertical circulation, and even if this has been fitted retroactively, its dimensions may not be compliant. The key reason for this is that Victorian or Edwardian buildings did not have the necessary lift technology to facilitate vertical circulation, and frankly, a central staircase was often the only option in most cases. To facilitate a retrospective lift installation to a pre-1920s property is also complex and potentially costly, which again needs to consider options to provide a reasonable adjustment.

Installing accessible toilet facilities to a pre-1920s building should be quite straight forward as this can be located in the same or similar location to existing sanitary rooms. However, an accessible toilet cubicle is larger than a standard toilet, and the repositioning of any existing internal walls will have to consider the structural impact. The actual fit out of the cubicle is standard and there is a uniformity in the design of these to be fully accessible.

Retail Construction and Characteristics

It is important to note that the term 'retail' is often associated with shops, but in the context of this book, the term retail is used to describe the commercial activity associated with the 'high street' such as but not limited to the following:

- Banks/Building Societies.
- Beauty Salons.
- Cafés/Coffee Shops.
- Charity Shops.
- Department Stores.
- Hairdressers.
- Out of Town Retail Parks.
- Pubs.
- Restaurants
- Retail Units.
- Shopping Centres.
- Supermarkets.
- Takeaways.
- Tattoo Parlours.

Most of these are quite straight forward low-rise buildings, which were purpose built or have been re-purposed between the different the different use classes. The retail sector is transient in the sense that it has constantly evolved or had to change since Victorian times. Having survived the town centre destruction afforded by the blitz in WW2 and the transition to out-of-town retail parks as well as shopping centres in the 1970s and 80s the biggest threat to the survival of retail and town centres has been the internet. The ability to shop online and have goods delivered to the door step has no doubt caused damage to the retail sector. However, the much-maligned *Death of the Highstreet* has resulted in a renaissance of experiential activities and use. While the impact online shopping has had on the traditional retail experience cannot be questioned, the consequence of the global Covid-19 pandemic in 2020 has also been profound. With a nation instructed to go into a lockdown for the best part of 2 years, the ability to work and shop remote proved the value of online access. However, emerging from lockdown saw a resurgence in the high street as a place to meet up and socialise. While the sales of goods appear to have moved significantly to an online resource, experiential activities that cannot be fulfilled online have returned to the high street. The further evolution and reinvention of retail properties has seen a substantial increase in all things experiential from bars, restaurants, hair and beauty through to tattoo parlours. It can be argued that at no other point in the 200 plus years of retail has there been a more appropriate time to afford access to goods and services.

The retail sector may be essentially divided into two distinct sub-sectors and these are:

- 'Traditional' High Street.
- Shopping Centres.

The sub-sectors differ considerably in the terms of location, size and construction technology, which has a bearing upon the consideration for the provision of an inclusive environment.

Historically, retail properties often are located in the centre of towns or cities and were the focal point for trading. This often dates back many centuries to the origins of a local market place, but in real terms, it was the Victorian era that introduced the essence of modern-day consumerism. Many high streets and town centres today are still littered with pre-1920s retail properties. Its testimony to the quality of their construction and versatility, combined with the evolution of planning laws which has resulted in their continued presence.

The architecture and building technology of traditional high street retail properties is essentially the same as the residential or domestic construction discussed when detailing low rise pre-1920s office accommodation. As a consequence of utilising shallow basements, strip foundations and load-bearing masonry walls, these are often low-rise properties restricted to three or four storeys above ground level. Despite this, it is relatively unusual for retail use in these properties to be situated above the second storey of such buildings.

Usually, the following age categories may be considered when analysing traditional high street retail properties:

- Pre-1920s.
- 1920–45.
- 1946–79.
- Post 1980.

Accordingly, these were noted to have had significant differences in construction technology. However, in the context of providing access to goods and services, they suffer largely the same problems. Parking provision in town or city centres is restricted with onsite parking to individual retail premises almost unheard of. There may be parking provision provided by street parking, local authority car parks or private car parking providers. The access routes to the main entrance are the same as low-rise town centre offices, and in many cases, the stepped front entrance presents immediate challenges for vertical circulation. Despite this, post-1995 re-purposing or shops that have undergone significant renovation are likely to have integrated solutions to navigate stepped entrances.

There is wide variety of entrance doors to commercial retail space and certainly with properties constructed or dated from the 19th century through towards the latter part of the 20th century. While a lack of uniformity in the principal entrances along with many individual shops fronts to retail may be seen to represent the unique nature of town centres, it does present access challenges. Assuming the entrance to a retail unit is step free, it is then necessary to establish if the door width or clearance is sufficient. Being able to pass through the door and open it are completely different actions, and accordingly, one of the significant challenges to an inclusive environment is the weight of the door opening. This equates to how much force is actually required to manually open it. Ideally, all main entrances to retail properties would not only have step-free access but also power-assisted entry doors. The counter argument is that this may in part contribute to generic high street entrances detracting from the individuality which contributes to their uniqueness. Accordingly, there are existing measures in place to conserve the historic built environment (as discussed in Chapter 8), and this is an example of the need to evoke design creativity in delivering a *reasonable adjustment*.

The internal horizontal circulation to retail properties is largely determined by the layout of the fixtures, fittings and equipment of the tenant or exploiter. In most cases, for retail shopping obstructions to horizontal circulation is created by the presence of display racks, goods and merchandise. For other occupation uses, such as restaurants, bars and cafes, the placement of furniture, partitions or screens will affect horizontal circulation in a similar manner to display racks in shops. Other more bespoke use of the space for hair, beauty or tattoos might require more specific fixtures, fittings and equipment that needs to be arranged to provide horizontal circulation, and it is accepted that if space is a premium and the equipment dimensions are critical, compromise might be necessary.

While it is accepted that most traditional retail space is situated on the ground floor and that these properties have been built mostly as low-rise three or four storeys buildings, some retail activity is undertaken on levels above the ground floor or even in basements. This introduces significant access requirements with a need to accommodate vertical circulation. This could be achieved with stair-lift access, and this would need to ensure any adaptation does not have a negative impact on any fire strategy, including the means of escape. The best way from an accessibility perspective to achieve vertical circulation is via a conventional passenger lift, although there is almost a complete absence of such installations to pre-1920s, and even those post-WW2. The construction technology of these largely residential type constructions simply does not have the

Figures 5.10 and 5.11 Retro installation of a platform lift to accommodate vertical circulation to a converted former bank into a coffee shop which is a listed building.

structural facility to support a lift facility. Post-1980s retail properties may have passenger lifts and certainly post-1995 new build retail properties or older properties renovated in the latter part of the 20th century to present day may have incorporated these. Modern lift technology has made it easier to retro fit lifts, but to most low-rise retail operators, the cost can be prohibitive (Fig 5.10).

Concerning the installation of sanitary rooms to retail premises, these are typically to those where the service provider is in the hospitality sector. Providing toilets are located in areas not compromised by the placement of furniture, fixtures and fittings restricting horizontal circulation or where vertical circulation is not an issue, then accessible toilet cubicles can be installed. However, their exact location may be determined by the availability of water supply as well as waste water evacuation connections.

Shopping Centres and Supermarkets

While the concept of shopping centres in the UK dates back to the city centre department stores of the early 20th century, which facilitated the complete shopping experience under one roof, the concept of shopping centres as known today dates from the 1960s. These were purpose-built facilities located away from town or city centres with good vehicle access. Their construction is largely quite simplistic utilising steel or concrete frames to offer good clear access between columns. There is much historical reference or suggestion to the word *Mall* or *shopping mall* emanating as a descriptive term used for an alley. Accordingly, this is believed to be the origins of *Pall Mall* in London. Irrespective of this, indoor shopping centres very much have the feel of streets enclosed and protected by a wraparound building envelope. In the terms of access, these can be potentially perfect to create an inclusive environment. Onsite parking facilities not limited by the existing built environment offer step-free access routes to the multiple entrances with wide unobstructive internal circulation permitting access to individual tenanted service providers. The nature in the uniformity of the building module or column grid is such that retail units are carved up in even dimensioned spaces separated by non-loadbearing internal walls. As a consequence of this simplistic but effective design, single units can be doubled or tripled in size be removing the internal walls. The secure nature of the shopping mall means that the external doors can be secured when the centre closes and therefore every individual retail unit can have fully glazed shop fronts without the risk of vandalism or break in so readily associated with town or city centre properties. Fully glazed sop fronts often translate into power-assisted entrance doors also with step free access.

Individual retail units and their placement of furniture, fixtures and fittings will determine horizontal circulation, and the uniformity of the infrastructure means typically the positioning of the services permits the effective placement of accessible sanitary fittings.

Shopping centres can be multi-storey but mostly occupy one or two levels over the ground floor. The large surface area of each floor plate means there are often multiple entrances, and accordingly, a need to install escalators and lifts. These are intrinsically expensive items of machinery to install as well as maintain. Multiple levels mean multiple escalators or increased lift capacity, and there appears a sweet spot concerning the floor plate size, number of retail units and levels or vertical circulation. The sanitary rooms and other facilities in shopping centres is usually very good with communal toilets

Figures 5.12 and 5.13 **Above and Figs 5.12 and 5.13: A purpose built and accessible, single storey supermarket.**

often located in one location which is accessible as are the accessible toilet cubicles. Also evident are sanitary rooms to some individual tenanted areas, but this will largely be dependent on the nature of the exploiter with hospitality use having their own compliant sanitary rooms.

In essence purpose-built shopping centres are likely to be the mostly compliant in the terms of providing an inclusive environment. In the terms of proving fully accessible shopping, there is only one more facility more accessible than shopping centres, and this is provided by online shopping. While online shopping and the internet are perceived to have contributed significantly to the death of the high street, it can be argued that this has had a more detrimental impact on shopping centres or out-of-town retail. Accordingly, many of these retail premises are facing the prospect of changing their use to offer experiential activities. While the consumer market will dictate the end use of these properties, the construction technology is such that they should always be adaptable to embody full accessibility.

Supermarkets comprise either large purpose-built facilities which are mostly single-storey offering similar inclusive traits associated with shopping centres, or these are more local convenience stores similar to traditional low-rise retail also with some of the associated problem areas for access. One common factor for supermarkets is that they rarely if at all occupy more than one level; accordingly, there are rarely issues concerned with vertical circulation (Fig 5.12).

Industrial Construction and Characteristics

Industrial buildings may be considered in two principal categories:

- Fabrication and Production.
- Logistics and Storage.

Factories manufacturing a wide range of products are often bespoke buildings that have been designed to perform unique and specific functions. While the building fabric may be relatively simple, it is the actual fit out and plant that make these highly unusual buildings. Many factories are owner occupied, and it is unusual for private investors to acquire these with a view to generating an income or return on their investment. The operation of manufacturing buildings is such that this is often divided into two key components that cover the prime function of manufacturing and the support facility directing operations or delivering customer support. Heavy manual or semi-automatic manufacturing is less likely to employ those with a physical or mental impairment; however, the operations and support facility may provide opportunities to employ those with disabilities. Accordingly, it will be necessary for employers and those providing a service or even access to goods and services to make the necessary *reasonable adjustment* to facilitate an inclusive environment. This in principle is a similar approach adopted by the logistics sector.

Logistics or storage buildings are something quite different to bespoke factories. These are essentially large, single storey, high-volume open-span buildings that historically were low-quality, low-lifecycle boxes that are strategically situated on the highways of the UK and Europe. The increased significance associated with online shopping has resulted in a rapid growth in the construction of these types of buildings in the early 21st century. Accordingly, modern logistics buildings are increasingly sophistic compared to those constructed prior to the online revolution. New build, high-tech logistics buildings come complete with the necessary support staff, and these are housed in some very high specification offices attached to the storage facilities.

The services provided by companies operating logistics sites encompass the movement of goods and loading these onto waiting hauliers in high-paced operating procedures where time is of the essence. For those on the shop floor, this is typically physical work and with the absence of available data concerning the quantity of employees who recognise as having a physical or mental impairment, it would appear that this sector has a low representation. The employment of those providing the vital management and admin support in the logistic process is more likely to be acceptable to those with disability. Accordingly, this facilities for both staff and those visiting logistics operations should be the same in principle as discussed for office buildings.

In summary, manufacturing and logistics buildings are usually located out of town or at city centres and often in bespoke industrial parks; invariably they are accessed by car, and in some cases, have poor public transport connections. The advantage of large private sites accessible car is that they usually have an abundance of parking space and, accordingly, there should be no excuse to ensuring that a sufficient number of these are accessible. The access routes from the car parking to the principal staff or visitors' entrances are usually level, and with these relatively modern buildings, the access should be step free. This differs from the vehicle access to loading bays or docks which are elevated for motorised access and accordingly do not need to be accessible to those with a physical or mental impairment. Older logistics or manufacturing buildings will typically have many of the similar access issues associated with pre-1995 buildings in the sense that these were not designed with access consideration. However, increasingly these are being renovated or upgraded for improved energy efficiency or sustainability; accordingly, this is also an ideal opportunity to upgrade accessibility. Older properties are likely to have manual access entrance doors, while power-assisted access doors may be more readily installed to the offices or modern logistics or manufacturing buildings (Fig 5.14).

Horizontal circulation to the offices attached to industrial sector buildings is limited to

Figures 5.14 and 5.15 Modern logistics buildings can have office construction to rival that of bespoke town or city centre office accommodation, there should be no reason why these buildings are not accessible to employees with a physical or mental impairment.

the placement of internal walls, partitions, furniture, fixtures and fittings. Older properties may encompass masonry internal walls that could also be loadbearing and accordingly could limit horizonal circulation. However, modern office buildings are likely to encompass non-loadbearing, lightweight internal walls that lend themselves well during the design stage as well as future renovation to deliver an inclusive environment.

Concerning vertical circulation to industrial sector buildings, this is likely to concern the accessibility of upper floors to the supporting office function. Most storage and logistics sites comprise a single storey to the warehouse facility, and while there may be mezzanine floor arrangement for storage or operation, these rarely utilise passenger lifts. Much of the internal operation of industrial sector buildings relates explicitly to the individual requirements of the occupier or exploiter (tenant). Therefore, much of the tenant installation includes potentially sophisticated fit-outs to accommodate the movement of goods or materials vertically in warehouses. This is different from the needs of office accommodation, and the need to ensure the workplace is accessible for staff or visitors to the building. While the office accommodation attached to logistics or manufacturing buildings may comprise a single storey, increasingly, these are becoming self-contained office buildings that are multi-storey in nature. Therefore, vertical circulation should be afforded with a lift installation, and this is becoming the norm to modern facilities; for older office accommodation there is often no lift facility. This is potentially complex and costly to fit retrospectively and is another example of the need to consider *reasonable adjustment*.

There is often a difference in the sanitary facilities provided within the goods side, production or warehouse of an industrial building. In line with the potential manual nature of work, these can sometimes comprise functional, hard-wearing, and easily cleaned finishes. While the manual nature of the work may prohibit the employment of staff with some physical impairment and hence the sanitary facilities are often not accessible, there is often sufficient space within the building to retro fit these. The toilet provision to any office accommodation supporting the logistics or manufacturing operation can vary significantly in the provision of an inclusive environment. Generally older, non-renovated buildings offer no or limited accessible toilets. As specifically with the office sector discussed previously, renovated or new buildings should be compliant and, in most cases, it is possible to retro fit accessible toilet facilities.

Residential

Certain large institutionalised investors do not appear to like residential investment as this seems to be an area of real estate investment that is just too 'personal'. Ultimately, residential investment deals with people and their everyday lives and may be considered to a have high hassle factor for the return on investment dealing with individual problems, complaints and issues.

There is a tendency to think about individual housing properties when considering the residential investment sector. This may be the case with local authority housing portfolios or even properties owned and leased at affordable rents by housing associations. However, it is also evident that investment properties are increasing likely to comprise large-scale apartment buildings with multiple living units. This may be the case with examples of residential investment properties including student accommodation or retirement apartments as well as more traditional apartments or dwelling houses.

Evolution in planning policies combined with a housing crises has resulted in increased obligations on developers to foresee residential units to be included in city centre redevelopment, and some of these to be classed as 'affordable' housing. Furthermore, changes to planning laws has also encouraged the renovation of office accommodation to include change of use to residential. This has had a big impact on the amount and quality of residential property coming on to the commercial real estate investment market, which may be considered financially appealing.

For some investors, there is an attraction in acquiring residential properties with some actively setting up management companies to undertake the property maintenance; this and perceived low risk income streams are the obvious benefit. Other investors are obliged to take residential investment properties as perhaps a 'trade off' for securing lucrative 'high end' office redevelopments. However, one observation may be that the tendency for increased mixed-use developments means that residential properties are likely to provide important facets of investment portfolios.

Contrasted with private sector residential investment is that of public sector social housing, which comprise a mixture of low, medium and high-rise tower blocks flats as well as large quantities of residential dwelling houses.

Social housing is a product of the welfare state which emerged after the WW2 when there was also housing crisis and a need to rebuild an estimated 450,000 damaged homes. Post-WW2 and with the birth of the modern movement, low-rise dwelling houses made way for a host of high-rise concrete towers with multiple flats occupying a relatively small building footprint. These were often situated in clusters with a hope to maintain existing communities. However, inadequate maintenance budgets and public-sector funding, allied with the evolution of concrete and design defects resulted in many of these residential areas becoming run down 'ghettos'. Moving full circle, many of these tower blocks have been heavily renovated or demolished and the sites redeveloped to include more low to medium-rise properties.

While there has been much criticism of the 'failure' of the high-rise social housing 'project', most public-sector low-rise housing portfolios comprise properties ranging from pre-1920s to post 1980s. While it may be suggested that generally the quality of materials used for most residential dwelling houses appears increasing lightweight with the passage of time, it should be noted that approximately 25% of the English housing stock (as per the 2015 survey) dates pre-1920s. Although pre-1920s properties have their own specific problems and pathology attributable to their construction technology, they do appear to have been constructed with good quality materials to a relative high standard. Consequently, significant quantities of public-sector low-rise housing stock dates pre-1920s, and therefore, the principal construction elements are of good quality. Indeed, government incentives with the 'right to buy' have encouraged home ownership for many social housing tenants. While this is a good thing, the negative is that insufficient public-sector housing has been constructed to replace it.

Irrespective of public or private sector residential development, it is logical to divide the property types into relevant categories. Naturally these should be either low rise (dwelling houses) or medium and high rise (towers). Concerning the implementation of an inclusive environment to residential properties, it is evident the there is a willingness for public sector owners to install access provision either on a case-by-case basis for individual houses or a more centralised approach for residential towers. A similar approach appears evident to residential properties owned and run by housing associations, although there appears no known obligation from planners to foresee specific number of accessible units per development. There is no obligation for private dwellings to be accessible although it is possible to adapt most residential properties to deliver an inclusive environment. However, with typical low-rise residential properties, there exists a need for design compromise which embodies *reasonable adjustment*. Most pre-1995 properties were simply not constructed with accessibility in mind and looking back through every previous age category, it can be stated that the relevant construction technology makes it increasingly difficult to deliver an inclusive environment.

Low-Rise Dwelling Houses

As discussed with low-rise retail properties, these may be divided into their relevant age categories ranging from pre-1920s to post-1980; there is a similar concept with in existence for residential property. Although there has been a slow progressive evolution in construction technology with advances in techniques and materials, the basic principles of the construction elements remain the same. In the terms of a commercial entity, low-rise residential properties are collectively grouped into the housing stock of local authorities or housing associations. To provide access solutions, it is necessary to understand that the properties included in the portfolios are classed as low-rise properties; therefore, these are typically single storey, two storey and often not higher than three or four storeys.

Many pre-1920s properties comprise town centre Victorian-terraced housing which are often robust red brick or natural stone two storey dwellings, often with no allocated parking provision and a principal front entrance. The nature of these properties is such that there is often an evenly spaced central passage ways which are typically gated and lead to a rear right of way or access route behind the properties. While there may be an access right of way, this by no means equates to a suitable disabled access route and does rely on free access between different plots. Pre-1920s properties typically utilise suspended internal timber ground floors, meaning there is often a raised step up to the front door or entrance porch. Most pre-1920s properties are set back from the road, street or pavement, and therefore, it is possible to undertake adaption to install a ramped access. However, where properties are located as such that their principal entrance is directly on the street or pavement, the retro fitting or access provision is considerably more difficult. Furthermore, some period property has a degree of grandeur with entrances raised significantly higher that a 150 mm step, and this is accordingly a difficult retro fit access.

The principal entrance doors to pre-1920s residential properties are not standardised, and the width as well as the weight of the door alongside the positioning of door furniture could be problematic for access. Furthermore, pre-1920s dwellings often have a passageway from the entrance leading to the foot of the staircase with doors through the internal loadbearing walls to a front reception room, rear dining room and galley kitchen to the rear. Toilets would have initially been situated externally in an outhouse to the rear of the property or centrally on the first floor. Increasingly toilets and additional bathrooms are located on the ground floor off or adjacent to the rear galley kitchen. In essence, pre-1920s residential properties are often of excellent original build quality, and many have withstood the test of time. Despite this, they can be restrictive in the terms of accessibility, and the need to potentially alter internal loadbearing walls means they can be relatively difficult to adapt to provide an inclusive environment.

There were quite a few technological advances in residential properties constructed between 1920 and 1945, which largely covers the time period of WW1 and WW2. These advances are profound in the sense that they addressed concerns over penetrating as well as rising damp, but this has had no effect of their ability for the adaptation to provide an inclusive environment. The period in history was important in the context of social housing as this is recognised as the time when the state started to clear poor quality slums and build replacements. Wounded and disabled soldiers returning from the front in 1918 were the subject of the 1919 Lloyd George election pledge to build *homes for heroes* but as revealed in the 1921 census released in 2021 was some apathy or scepticism towards this slogan. As often is the case with disability, actions speak louder than words. It is easy to be cynical in the 21st century as even despite the introduction of legislation there appears more said than done regarding disability. Despite this, there was some attempt at designing and building residential properties specifically for soldiers returning from war with disability, as discussed in Chapter 3. Generally, most houses constructed post-1920 adopted different aesthetics when compared to the former traditional Victorian terraced properties.

One of the defining features of such houses is that they often had outdoor space including front gardens, back gardens and a driveway. Furthermore, the style of houses meant that detached and semi-detached properties became popular, as well as the introduction of the concept of single storey living known as bungalows. These features have contributed significantly to social housing and in particular the ability to provide accessible parking spaces adjacent to properties. The obvious advantage of such housing is such that these spacious and well-located properties were attractive propositions. These along with many early to late 20th century properties were the subject of the *right to buy* scheme in the 1980s, meaning their ownership became private and no longer formed part of current local authority housing stocks.

Post-1920 and particularly after WW2 and what is considered by some as the birth of the welfare state meant there was an attempt to build multi-storey residential flats or apartments. This in part was attributable a significant investment in social welfare with the birth of the National Health Service (NHS) in 1948; this had a focus to re-build a country ravished by war. A post-WW2 shortage of timber and other building materials was in part responsible for innovation in construction which saw the use of concrete for the construction of residential properties including the structural frames of early multi-storey properties. Typically, four or five storey concrete-framed residential apartment buildings consisted of simple dwellings connected by communal walkways and staircases. There is no doubt that designing for disability was something not considered in these early muti-storey properties, and one obvious concern is the likely absence of lifts to provide appropriate vertical circulation. Despite this, the majority of the individual flats or apartments were arranged over one level and provided vertical access to the entry level is achievable the there is no need for additional internal vertical circulation to be provided. The reality is that many post-WW2 city centre apartment buildings have been the subject of renovation or redevelopment, and assuming this was done towards the latter part of the 20th century, then it is likely that the provision for disabled access should be foreseen with public sector housing.

In the aftermath of WW2 and instigation of the welfare state, there was an unprecedented quantity of housebuilding in the UK, and it was primarily led by public sector, as illustrated in data published by Statistica[1]. The data illustrates as surge in local authority house building in the mid-20th century followed be a rapid drop off as politically and economically the country adopted a more capitalist approach (Fig 5.16).

Significantly in the terms of providing disabled access in the 21st Century, local authority and housing association accommodation are more likely to facilitate this than private residential developments or the private rental sector. Despite this, there are a number of adaptations or design solutions that can be made to private dwellings that may be funded privately or with local authority support. As indicated in the published data,

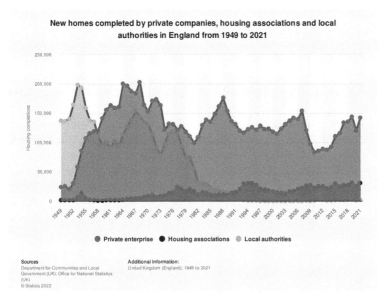

Figure 5.16 New homes completions 1949–2021.

the majority of public sector housing was constructed immediately after WW2, this gradually receded but declined significantly in the 1980s in line with the right-to-buy scheme. Despite the pledge to replace public sector housing sold off for private ownership, the gap has been in part plugged by housing associations.

Prior to the eventual and terminal decline in the building of public sector housing there was a brief peak in the 1960s and 1970s, which was when most of the inner-city high rise residential development occurred in the UK.

Medium- and High-Rise Residential Towers

There are many differences between low-rise and medium or high residential investment properties, and it is recognised that mass constructing multi-storey residential buildings occurred from the 1950s onward. The principal observation concerns the structure, and this will be either reinforced concrete, steel or a combination of both materials. The structural grid will have been designed to accommodate the required window/facade modules for the individual residential units. These is constructed around a central reinforced concrete core used to house lifts, services shafts and internal emergency staircases. High-rise residential development pre-1990 paid little attention to the provision of disabled access; however, many of the developments constructed in the 1960s or 1970s have either been significantly renovated or demolished and redeveloped. In this instance, there should be little excuse as to why these do not encompass disabled access. Despite this, it should be recognised that it is very difficult enlarge or widen existing reinforced concrete lift shafts in the event that larger lift cars are required to facilitate accessibility.

Many residential towers constructed in the 1960s and 70s were executed using exposed reinforced concrete panels and structure. This has left a legacy of defects and material failures which has had, and will continue to have, significant impact on investment in these types of property. Latterly, it may be observed that façade cladding systems to modern (post-2000) residential towers appear to have become more lightweight with increased levels of thermal insulation. However, these materials have also their own specific building pathology, which is starting to become evident. Noticeable is a major concern with the use of Aluminium Composite Materials (ACMs), which presents a major fire risk as exposed by The Grenfell Tower fire in 2017. While this does not affect the provision of accessibility, it is an example of retro fit in search of a building improvement, which has caused a significant issue. As with most retro fitting of building adaptions, improvements or attempts at compliance, these are often limited or restricted by the existing building features. As a consequence, retro fitting is often more costly than when undertaken as part of the initial design. However, this should in part only suffice as a limited excuse by virtue that service providers are obliged in law to undertake *reasonable adjustment*. Perhaps one of the biggest restrictions or limitations on the retro fitting of public sector housing is the financial resources of these organisations.

With limited available funding for public sector housing and the need to provide accessibility, there is no doubt a challenge concerning vertical circulation. This will need to be provided by lift installations, which are notoriously expensive to construct, maintain and operate. When such installations are out of service, then this poses major accessible issue for both those with or without a physical and mental impairment. While many of these developments have external areas accessible at street level, there was historically a design consideration to connect residential tower blocks with walkways, landings or steps, which are often not step free. Furthermore, the use of both cast in situ and precast concrete internal walls places potential structural limitations on internal circulation, such as door widths or the placement of accessible bathrooms. These structural limitations mean retro fitting of accessible facilities are likely to be harder to implement.

While there has been a significant dropoff in new build social housing developments post-1980s, there has been an increase in the presence of housing associations who operate commercially as property owners or investors in the delivery of affordable housing schemes. Those housing associations operating post-1995 should be obliged to ensure that appropriate access provision is foreseen or the necessary adaptations are made in their capacity as service providers. This will be easier to achieve with new build or relatively modern properties but as previously alluded to, more complex with older housing stock.

Student Accommodation

While it has been established that buy to let properties has been seen by private landlords as a way to generate very effective income streams, increasingly, commercial investors are investing residential projects aimed at the private rental sector. One example of this is student accommodation which is a growth area of commercial residential provision. As service providers, these private organisations are obliged under the Equality Act 2010 to facilitate accessible accommodation. There is no doubt some relatively old student accommodation, or town houses and properties located within historic buildings that may not be readily adapted to provide an inclusive environment.

Post-1992 there has been a significant expansion in the provision of higher education in the UK, accordingly the numbers of academic institutions offering degree courses has expanded and with this, the numbers of student accommodation. Therefore, there has been widespread new build and renovation of university campus accommodation along with a wide range of private investors constructing student housing in university cities and towns.

Much of these properties will have been developed to applicable standards and regulations, therefore provision for disabled students should be foreseen. Indeed, with universities competing for students on a number of levels, many are keen to illustrate the inclusivity of their academic courses and provision. Accordingly, many universities openly advertise on the web pages or prospectuses the measures afforded to accommodate those with a physical or mental impairment. In the terms of the need to provide parking, adequate access routes, main entrances, as well as vertical and horizontal circulation, there should be no reason why modern or renovated buildings should not be able provide this. With the widespread availability of accessible sanitary rooms on campuses, adapted or bespoke student accommodation should have accessible toilets and wet rooms.

Care Homes and Retirement Living

While most commercial providers of residential accommodation seek to make a relatively small percentage of their service provision accessible, there is one growth

Figures 5.17 and 5.18 **Above (left): Over 55's retirement apartments and (right) Fig 5.18 A residential care home.**

area where accessibility is at the forefront of design. This is within the provision of care homes for the elderly, which can also encompass assisted living and relative high-end residential accommodation associated with *over 55s* housing. As these properties cover a wide range of ages for the occupants, who will continue to age through their residency; it makes sound business sense to make these properties as inclusive as possible.

The correlation between age and disability has been long established and is discussed extensively in Chapters 3 and 7, and it is in the interest of both the public and private sector delivering retirement living or care homes to provide an inclusive environment. While it is recognised that there will be some older or even historic buildings providing this service, a bit like the provision of student accommodation, this is a growth sector. Accordingly, there are a number of private investors and organisations developing these commercial buildings and there should be nothing commercially preventing full access provision. As with the delivery of an inclusive environment within the built environment, it will always be easier to achieve this with a new build project over a renovation to accommodate all the access requirements previously listed (Fig 5.17).

Leisure

The leisure sector comprises a large variety of potential investment and owner occupier properties, these may be wide ranging from sports venues and stadia through to concert halls, entertainment venues, museums, restaurants, cafes, bars, gyms or holiday complexes, hotels and guest houses. The principal observation with properties within the leisure sector is that these are used to house staff who work in them but also relatively large amounts of transient visitors. Unlike the other building sectors, the majority of people occupying or using these do so by the hour, day or week; they are rarely occupied by permanent long-term residents. The service providers operating in the leisure sector are varied, and accordingly, while some might offer generic services such as hotel accommodation or provide restaurant facilities, there are some highly bespoke services operating out of unique buildings.

In accordance with The Equality Act 2010, there is an obligation placed on service providers not to discriminate on the basis of disability and accordingly, it is recognised that there is a requirement to make reasonable adjustment to facilitate access to goods, services or employment.

Hotels, Pubs and Restaurants

The most common type of leisure investment for property funds are hotels, but this sector is also occupied by lots of specific chains as well as independent owner-occupied businesses. Many large and small hotel premises also house bars, restaurants and cafes as well as potentially meeting rooms, conference facilities and gym facilities as well as swimming pools, in some cases. In the UK, there are also tens of thousands of pubs located in most areas of the country. While many hotels are often situated in town or city centres, it should be recognised that there is also a large quantity of properties, particularly pubs which are located rurally and form the focal point for many communities. Buildings in this sub-sector may be owner occupied or leased and fit out to the requirement of specific brands or chains.

Irrespective of the hotel rating or brand and also concerning pubs, restaurants, and cafes, from a technical perspective, it is important to sub-divide these buildings into those that are purpose built and those that have evolved through with extension, renovation or conversion. The reality is that any new build, conversion or renovation undertaken post-1995 should have taken inclusivity or 'disabled access' into consideration primarily due to The Disability Discrimination Act 1995 and compliance with building regulations.

Despite this, there is no doubt many hotels are historic buildings with a wealth of important historical context as well as materials. These are certainly likely to originate from the pre-1920s or, in many cases, pre-Victorian or Georgian. The older the property, the more likely it is to be listed or located within a conservation area; these are significant factors to consider when attempting to undertake reasonable adjustment for the provision of an inclusive environment. Concerning the provision of access to goods and services as well as facilitating an inclusive environment for those working in these establishments, there will always be challenges associated with horizontal and vertical circulation. One consequence of older historic properties in this sub-sector is that they are likely to be low-rise properties with occupation rarely being four floors above ground level. While it might be considered relatively straightforward to negate an entrance step with a permanent or temporary ramp, the alteration of internal walls to accommodate horizontal circulation may be problematic due to their loadbearing nature irrespective of their potential listing. Likewise, these older and often historically significant properties will have almost no possibility to install a lift shaft or lift capable of achieving vertical circulation. It can be argued that adapting entrances or possibly providing a secondary step free entrance can often be achieved hence entry to these buildings can often be afforded. Chapters 8 and 9 consider some of the ways to achieve design solution and reasonable adjustment, but in some cases, the geographical location of some of these premises may render them impossible to access if there is simply little or no access route foreseen. Not all historic hotels, restaurants or pubs have their own dedicated parking and access routes to the principal entrances may often be governed by local authority roads or pavements. Overall, this sub-sector along with similar issues associated with historic museums or places of worship is probably one of the most challenging with respect to the provision of an inclusive environment.

In contrast many hotels dating post World War II are likely to be purpose-built leisure properties. With advances in construction technology, these will inevitably be concrete or steel framed, and although also located on confined sites in town or city centres, these are often medium or high-rise buildings. 'Modern' hotel properties encompass the same or similar structural design to office or apartment buildings with internal separating walls also being non-loadbearing. This type of construction including the retro fitting out to facilitate access will always be less complicated on the basis that internal walls and partitions can be arranged to meet the necessary dimensional requirements (Fig 5.19).

Holiday complexes are a concept essentially emanating post WW2, and most of those constructed in the 1960s to 1990s are likely to have undergone significant upgrade, renovation or redevelopment. Accordingly, the buildings and services provided should be fully accessible with fewer restrictions typically presented by older, historic properties. Many of these facilities are low rise and often single storey, which negates the cost investment required to accommodate vertical circulation. In the event that there is a need to access upper floors, it may be commented that the provision of lifts and all other accessibility features are a reasonable adjustment and expectation.

Figures 5.19–5.22 Top left: (Fig 5.19) A Grade 2 listed hotel originally dating 1863–68 with a sympathetic attempt to accommodate an accessible ramp although the gradient and detailed does not appear to meet the necessary design criteria (see Chapter 9). Top right: (Fig 5.20) A Grade 2 listed pub reported to originally date 1636, which by the very nature of the date and listing may experience difficulties with accessibility. Bottom: (Figs 5.21 and 5.22) A modern bespoke hotel where accessibility has been included in the design and not as a retrospective alteration.

Sports Stadia, Concert Venues, Theatres and Cinemas

The rise and prominence of the Paralympic movement has showcased the participation of those with a physical or mental impairment at the very level of competitive sport. It is important recognise the significant role that the Paralympics has had on the recognition, as well as promotion of disability rights; this is covered in Chapters 2 and 7. This book is concerned more with the actual accessibility of the buildings and venues hosting sports from a participation and spectator perspective and not the actual inclusivity of the sports themselves. The need to access sports stadia and other spectator venues such as concert venues and theatres will often be dependent upon the age and location of the property. Most buildings operated post-1995 should be inclusive and as will all of the buildings associated with commercial operation should consider parking provision and access from this or the public domain to the main entrance along with both horizontal and vertical circulation with specific access provision for toilets, facilities and services.

New and bespoke facilities should be fully accessible, but it is a widespread notion that there are often not enough accessible seating areas including wheelchair spaces and the seating for carers. A number of high-profile sporting or music events have come under criticism for simply not providing sufficient numbers or accessible seating areas or for not permitting guide dogs or another accessible requirement. Compounded to this are the 'products' of the performers which are not always inclusive, and increasingly, there is a recognition to provide sign language and audio description at these events.

While new purpose-built venues can be wholly inclusive, there are a number of historic venues with listed status, thus protecting their architectural and cultural past. As detailed in Chapter 8, it is possible to provide access to historic buildings, but under the notion of 'reasonable adjustment', this will involve compromise. As discussed later, when considering access to the 'historic' build environment, these buildings are not always the classic Victorian or late 19th or early 20th century buildings expected. Fine examples of post-WW2 or modern architecture is sometimes given listed status by virtue of the fact that these are mainly concrete heavy buildings. These are examples of pioneering architecture reflecting a new post-war freedom in expression, culture and society. In reality, many of the buildings and their designs have been plagued with defects and problems; however, despite much perceived failings with the architecture, society has come to respect and even admire or treasure this architecture. Such examples of renovated public auditoria and concert venues exist in several cities of the UK with the *Southbank Centre* in London often held up as a fine example of it. Although not officially listed and indeed given a certificate of immunity from listing in 2020 to 2025, this much-loved 1960s concert venue and collection of cafes, bars and restaurants has attempted to remove as many barriers as possible to facilitate access (Fig 5.23).

Retro fitting for an inclusive environment with sports stadia, concert venues, theatres and cinemas is always going to be more problematic and potentially more costly when compared to modern purpose built premises. However, there are sufficient examples of where this has been done, although the greater question is to define what is a 'reasonable adjustment'. While this is intrinsically discussed in several chapters of this book, there needs to be a recognition that there is a complexity in the individual disability specific requirements for the access of goods, services and employment. Defining and implementing a uniform approach to reasonable adjustment has been made through academic

Figure 5.23 **Accessible spectator facilities at a renovated sporting arena with space for both wheelchair users and companions or carers.**

research. However, this needs to be road tested and potentially tweaked according to the experiential findings of those with a physical or mental impairment.

Museums and Tourist Attractions

While under The Equality Act 2010, it is not permitted to discriminate against those with a physical or mental impairment with regard to providing access to goods and services, it is often the buildings that provide the biggest barrier to inclusivity. Museums and tourist attractions can provide both internal and/or external experience for visitors. Some of these are standalone sites in acres of grounds and others in towns or cities. The default assessment in the provision of access or the need to undertake reasonable adjustment will look at the presence of onsite parking, disabled dropoff points, as well as the principal building entrance. Clearly, there are some attractions thousands of years old, such as Stonehenge or even those a mere hundred years or so old which were clearly not initially designed with inclusivity in mind. However, where more recent visitor centres have been constructed or parking has been added and foreseen, then there is no reason why this cannot be deemed accessible. More complicated are town or city centre museums and tourist attractions were parking and the immediate access route is likely to be local authority owned.

The actual access into buildings which may be listed or within conservation areas are subject of the inevitable reasonable adjustment, which can often be done to facilitate this. The internal dimensions of museums and tourist attractions often include a contradiction and myriad of different measurement, clearance widths as well as difficult changes in levels. While much of this is discussed in Chapters 8 and 9, reasonable adjustment is not always possible or indeed reasonable. The cost of vertical circulation

Figure 5.24 **A modern bespoke museum and it is inexcusable that this could not be fully accessible,**
however, it is necessary to consider that many museums reside within or are historical
buildings.

and the provision of lifts are technical complex to execute, accordingly the 'reasonableness' may come down to the physical restraints of the existing buildings or the feasibility
in cost.

Because of the uniqueness of many tourist attractions and despite the obligation to
undertake reasonable adjustment, there does appear the adoption of the *path of least
resistance* by service providers regarding attempts at providing an inclusive environment.
Too often, the term reasonable refers to what is feasible to the building owner or service
provider with less focus on the actual provision of a tangible attempt at providing access.
There appears a willingness for some to hide behind the protection afforded to the listed
historic built environment to argue against implementing alteration. Notwithstanding
this, there are a number of individual attractions as well as large organisations with big
portfolios of historic attractions that actively recognise the need to deliver an inclusive
environment; this is discussed in significant detail in Chapter 8 (Fig 5.24).

Public Sector

Investment in public sector properties is either derived from tax revenues or via private
investors providing investment or properties that are effectively leased back solely for
public sector use. The public sector is vast and includes some of the following key sub-
sectors:

- Healthcare.
- Social Welfare & Housing.

- Education.
- Defence.
- Criminal Justice.
- Local & National Government.
- Infrastructure.

Within the public sector, there is a wide and diverse building stock with many bespoke and historically important properties. As with all other building sectors, it is important to note that the provision of an inclusive environment will depend largely on whether the buildings were existing pre-1995 or whether these have been the subject of renovation, extension or are newly built properties from the early 21st century. The defining principle of providing access to goods, services and employment for those with a physical or mental impairment remains that it is always more complex and costly to do this retrospectively. Despite this, it should be stated that buildings owned or operated by the public sector or buildings fulfilling a public function, such as railways stations and other transport hubs, generally have good access provision. While there appears quite a lot of criticism directed to the general provision of public transport, there is a notion that this is magnified for those with a physical or mental impairment. It is important to recognise the inadequacies of public transport provision such as spaces on buses or trains, which have been the subject of much criticism. This book concerns primarily the physical transport hubs and buildings that provide a vital part of public transport provision but also recognises that transport companies should be doing more.

Historically, there has been a great sense of national pride concerning infrastructure projects or public-sector buildings. With the industrial revolution and the associated advances in architectural and engineering design capabilities, high-quality construction projects were conceived and executed. There are in existence many examples of Georgian, Victorian and early 20th century public sector or national infrastructure projects today. These include:

- Hospitals.
- Schools.
- Courts.
- Prisons.
- Town Halls.
- Railway Stations.
- Bus Stations.
- Libraries.
- Leisure Centres.

Some of these are now under private-sector ownership or are enveloped in private sector partnerships for funding and operation of these vital assets. Historically, public sector buildings were owner occupied and accordingly were often built with high-quality materials by skilled tradesmen. There are many examples of high-quality Victorian public sector buildings that epitomise the sense of pride and civic status that they were built with over 120 years ago. This sense of civic pride appears also to have been evident in the early to mid-20th century following the end of WW2 and the rebuilding of a country ravished by conflict. The birth of the welfare state and the NHS in 1948 exemplified the importance of the public sector with the construction of numerous public-sector projects. It can be argued that towards the end of the 20th century, there was a political shift towards a more capitalist and less socialist model of government, with all political parties adopting a centre-right approach. Lower levels of taxation have contributed in part to reduced public-sector funding, and the influence of private sector investment in public-sector projects.

Despite the shift in funding of public-sector projects, it may be commented that the majority of public-sector buildings or those providing public service generally have made consideration to access provision. Indeed, the public sector appears outwardly to support diversity and inclusion, this is evident from an employability and access perspective.

The unifying principles of access provision to public-sector buildings or those facilities providing public service are the same as all other commercial property sectors. This always revolves around parking provision, the number and quality of accessible spaces where relevant through to the principal access routes, main entrance, horizontal and vertical circulation as well as sanitary provision. The individual accessible service provision will depend upon the function of the organisation residing in the buildings.

Hospitals and Health Care

As perhaps expected, the nature of hospitals is such that these are amongst the most accessible buildings within the commercial property sectors. In essence, it was the Victorians who were pioneers in the provision of care for the sick and infirm. Looking back, it appears an obvious curiosity to examine the ailments of the human body with some unusual and unique diagnosis. But it appears highly inappropriate to scoff at early medicine when there is a sense that this has always been concerned with patient and outcomes. The construction of the buildings housing these revolutionary practices were grand in design, reflecting both the pride and power embodied by the medical profession. While some of these buildings are still in existence today, it is apparent that they were no originally fit for purpose in the context of modern approaches to accessibility.

Traditionally, hospitals and medical practices or GP surgeries were located centrally in cities or towns in close proximity to the communities they served. While the sites might be considered tight and restrictive by modern bespoke-built hospital facilities, many of these have had sufficient land to accommodate full or partial redevelopment. Much of the historical architecture associated with the medical profession has been in part listed or protected. The reality is that the original sections of the buildings are often only part of a hospital complex, despite performing the forward-facing façade of gateway to the service with a complex warren of corridors, wings and wards built off these in sections representing the decades of following medial progress (Fig 5.25).

Apart from the obvious problem of parking provision in city centre locations, there is an immediate requirement for hospitals to have well-signed, well-lit and accessible access routes leading to step-free entrances. While this might be difficult to achieve with historic or listed buildings, there is almost always no problem with accessibility. The need to transport internally patients, medical equipment and staff efficiently means that both horizontal and vertical circulation are facilitated. Furthermore, the placement and conformity of sanitary provision, seating, counters and services are usually more than adequate. The nature of the service provision is such that there are competent and trained

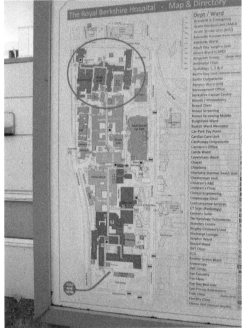

(caption on next page)

Figures 5.25–5.29 Top left: (Figs 5.25 and 5.26) The 'Royal Mineral Water Hospital' dated 1739, which despite caring for the sick and infirm was not designed or built with accessible considerations. Middle left: (Fig 5.27) The Royal Berkshire Hospital with the original section (top/purple) dating pre 1920 and Fig 5.28 (middle right) but as with many public sector health buildings, added to and extended with more modern sections (bottom right – Fig 5.29) post 1920 to present day.

medical staff, as well as porters who are able to transport those with physical and medical impairments effectively through these buildings.

Increasingly, city centre hospital sites and smaller, historic 'cottage' hospitals are being closed for cost-saving and efficiency reasons or due to the value of the site for redevelopment. As such, there are increasingly a number of large primary care doctors' surgeries and new hospital facilities being constructed on strategic road networks or close to ring roads on bespoke greenfield sites or redeveloped brown field locations. Clearly there is no reason why these modern buildings should not be fully accessible. Notwithstanding this, there still remain a number of medical practices including doctors, dentist, physiotherapists and other services located within the centres of towns and cities with quite poor access provision. The obvious absence of onsite parking or dropoff combined with often 'residential type' construction technology means that horizontal circulation may be restricted as well as almost no vertical circulation provision. Other than situating services on the ground floor and negating potential steps from street level to the main entrance as well as reducing the need for too much horizontal circulation, there are challenges to providing an inclusive environment.

Schools

Similar to the health care sector and inline with much of the public sector buildings delivering goods, services as well as employability, schools are either historic buildings, purpose-built academic institutions or the former which has been the subject of redevelopment, renovation and extension.

With many older school buildings, the reoccurring problem of vertical as well as horizontal circulation can prove troublesome. There is a relatively low percentage of children recognising as having a physical or mental impairment within the population as a whole. However, despite this, the number of children with mobility requirements is significant when it comes to recognising the need to navigate the obvious obstacle presented by a lack of vertical circulation. As a service provider, the installation of a lift will present on of the biggest investments to accommodate an inclusive environment. This will be easier to install for bespoke modern education buildings when compared to retro fitting older properties. While integrating children with disability into mainstream education opens up the full national curriculum and offers the opportunity to mix with able-bodied classmates, the charity Scope suggests that what is right for disabled children will depend on their needs and personality.[2] Accordingly, the alternative to mainstream education are specialist schools, and it is likely that these are not only better equipped and resourced for teaching children with a physical or mental impairment, but also are designed for full accessibility.

Courts, Town Halls and Libraries

Courts, town halls and libraries often occupy the very centre of communities, and while court buildings may be located in larger provincial towns or cities, town halls and libraries are much more common. Historically, these are often situated in historic and significant premises, but despite this, there is often widespread evidence of reasonable adjustment to facilitate access. Despite being the subject of public funding, there appears a genuine willingness to accommodate disabled access; this is in stark contrast to commercial real estate investment, which appears obsessed with compliance to legal minimums.

The central locations of these civic buildings are such that there is often no or minimal parking facilities other than public car parks or those operated by private parking companies. There are numerous examples of good quality privately owned and operated car parks where accessibility is seen as an essential commercial requirement. In contrast, it should be noted that the car parking provision in smaller, historic towns may not only be limited in number but also technically non-compliant in dimensions, signage etc. The suitability of the access routes from public or private car parking facilities will inevitably depend upon that provided by the local authority. While a service provider is obliged to make reasonable adjustment to facilitate access, this is not the case with access routes.

Despite often being located in older or historically important buildings, civic buildings often have examples of effective reasonable adjustment to the main entrances as well as accommodating vertical circulation. The notion of horizontal circulation is dependent upon typically the clear widths of internal corridors and, to older properties, many of the internal walls are loadbearing, so removal or widening of these could be problematic.

Figure 5.30 A pre 1920s single storey, public library with a level approach and automatic, power-assisted access door.

Modern buildings fulfilling a civic function are likely to embody concrete or steel frames, and accordingly, the internal walls are not loadbearing. While the location or placement of the supporting structural framed cannot be readily altered without potentially compromising the stability, non-loadbearing internal walls can be moved to widen corridors and facilitate horizontal circulation (Fig 5.30).

Vertical circulation within civic buildings invariably depends on the presence and compliance of an appropriate lift installation. While once more this should not be an issue of concern with modern buildings, it will be a more complex issue with those that are older or historic. Despite this there are often attempts within these public sector buildings to accommodate vertical circulation and provide an inclusive environment.

The provision of accessible sanitary installations is often foreseen within the public sector and even a retrospective fit out should be possible. Ultimately the space required to install an accessibility compliant cubicle is not significantly greater than an existing cubicle or it may be possible to replace two non-accessible cubicles with one that is fully compliant. The technicalities of this are discussed in greater detail in Chapters 8 and 9.

Public Transport Hubs

Public transport hubs are essentially the buildings that are used for travel; these can be both railway and bus stations as well as ferry terminals and airports. These can be large and complex buildings that have high volumes of passengers, but transport hubs are not considered to be individual pickup or dropoff points on bus routes. There is a recognition that to access buildings, there is a key requirement to first travel to these destinations. One observation is that bus stops and pedestrian routes within the built environment are simply not good enough to engender true inclusivity. The responsibility of this lies with the local authorities, and there are supposedly guidelines or best practice concerned with providing appropriate wayfinding. Despite this size, and although not acceptable from an inclusivity perspective, the sheer quantity of linear metres of pavements to town and city centres appears unmanageable from a maintenance perspective, let alone to consider accessibility.

While key transport hubs appear accessible, some are limited through their age; one example of this is the London Underground, which has some access restrictions attributable to Victorian tunnel networks, often with several level changes. Despite this, a number of underground stations are advertised as being accessible on the Transport for London (TFL) tube map. The underground network solely concerns London, and despite the array of historic tunnels as well as subterranean platforms, may of the stations above ground are accessible. In some instances, stations are listed, and accordingly, all of the compromises associated with access of the historic-built environment are evident. The underground is just one mode of transport, and there is the presence of a UK wide train network providing vital transport facilities that needs to be considered. Train services and access to stations are managed by either Network Rail or the individual rail companies operating passenger services. Clearly accessibility on the trains themselves is a priority for disabled passengers, and this book is not concerned with the role these organisations play as service providers. It does, however, concentrate on the built environment and the physical buildings that constitute the transport hubs. Much of the UK train network originates to the Victorian era and granted is the many decades of investment which has been made to include the upgrade of these including funding for accessibility.

Under the Equality Act 2010 and the obligation for service providers to undertake reasonable adjustment to facilitate access, there appears a commercial recognition of the term 'reasonable adjustment'. While it is apparent that access provision is evident to many mainline stations and those in areas of relative high-density population, a number of stations are not accessible. A shift towards electronic tickets and trains without guards means increasingly that remote stations are not manned; furthermore, outside of key commuting hours, even significant stations are often devoid of staff. This is important as disabled travelers often require assistance, and such absence prohibits their ability to carry out normal day-to-day activities. It is not just about providing staff to man stations but also the physical characteristics of the station buildings that pose an obstacle irrespective of the presence of staff. Clearly there is a need to access station entrances and concourses to access ticketing facilities. Considering the notion that this can be provided electronically, it is feasible that passengers, irrespective of disability, can obtain tickets to travel, this nonsensical approach introduces discrimination. The single biggest barrier is the absence of step-free access to platforms, which usually require foot bridges passing over train lines. This is likely to prove problematic to older historic stations or those in relative remote locations. Here the crossing over tracks may be via a footbridge with no lift or by physically crossing the rails on a designated level crossing, which are often not manned. In the context of a 'reasonable adjustment', it could be argued from a commercial perspective to install, operate and maintain a lift for remote railway stations would not be reasonable. This is where the term 'reasonable' is utilised in a subjective capacity. There is however an honesty-with-access provision to the rail network buildings in the context that information on accessibility and in particular step-free access is publicly available on the web pages of *National Rail Enquiries*. If anything, this serves a purpose to allow for the planning of journeys, but in particular, creates a list of destinations that cannot be visited due to the lack of access. In the context of current society and a seeming willingness to provide access, this is a reminder of the plain facts around the social model of disability and that it is the lack of access provision which propagates this.

If the location of a station or historic importance of the building is such that it is not accessible through barriers in accommodating vertical circulation, it is also likely that these do not have accessible sanitary facilities. Despite this, there are many examples of stations and other transport hub buildings where accessible toilet facilities are present. Provided there is available space for what are relatively small dimensions to fit able bodied toilet cubicles, there should be space for accessible sanitary rooms.

Leisure Centres

The concept of leisure centres or places of exercise is not wholly new and public swimming baths were initially constructed in the 19th century. Indeed, many of these buildings and the elaborate outdoor lidos of the early 20th century are listed. This potentially presents obstacles for access in much the same way as has been discussed about historic commercial buildings throughout this chapter. Despite this, the rapid rise in leisure centre provision towards the end of the 20th century has meant that there are a large number of both privately and publicly operated facilities. As expected, those post-dated 1995 are likely to be designed with accessibility encompassed in the design. Most of those pre-dating this period have probably been the subject of renovation, redevelopment or the subject of retro fitting for reasonable adjustment.

The design of leisure centres is such that these are often single storey buildings, but sometimes encompass a mezzanine level or first floor but rarely exceed two storeys in height. While ground floors are often used to facilitate the locating of main sports halls, changing facilities and swimming pools, the upper floor may be dedicated to fitness studios and gym space. The placement of lifts as original features or retrospectively may be considered relatively straightforward to these largely open plan buildings with supporting structural frame.

In most cases, there is onsite parking provision, and this can also easily be foreseen as having a number of accessible car parking spaces. Likewise, the access routes to the main entrances of these building types are wide, even, and with minimal gradient. Most modern or renovated buildings have access entrances with power-assisted doors leading to open plan reception areas with accessible entrance desks. The presence of non-loadbearing internal walls means that horizontal circulation is rarely restricted by the presence of structural components. Often changing rooms and sanitary facilities are adequately provisioned, and as these are enclosed with non-loadbearing internal walls, the placement, adaptation or even repurposing of the space to ensure inclusivity is not technically complex.

In essence, as leisure centres are both relatively modern buildings and often operated by local authorities of in partnership with the public sector, these typically have a high level of accessibility.

Summary

Commercial property has historically been seen as a safe investment vehicle for private investors or is used facilitate the provision of public-sector civic function. The occupiers or tenants who reside in these buildings are obliged under The Equality Act 2010 to undertake reasonable adjustment to facilitate access to their goods and services. Individual service providers and employers will have to decide if the prime function of their business operation is inclusive, and this might be, for example, the production from a theatre company; however, the other critical access requirement is the theatre itself. While the disabled user may look at the service provided and access to the building as a complete entity or experience, the two things are entirely different. This text is concerned with the physical aspects of accessing buildings for goods, services or employment.

With respect to the accessibility of the buildings housing goods, services and employment, it is necessary to apply a relatively simple criteria irrespective of the building type or destination use. While the destination use has a significance on the design of buildings and their layout, which can affect their accessibility, building age as well as location potentially represents the greatest access barrier. Despite evidence that some sectors of society have historically provided help and support to those with disability dating back to medieval times, incorporating inclusive design is a relatively new concept. Driven by legislation and regulations as well as the threat of prosecution for noncompliance, accessible design was only enforced in the latter part of the 20th century. As a consequence, the vast majority of existing buildings pre-date the legislation, resulting in a raft of compromise associated with retrospective adaption for inclusivity emanating from the obligation to undertake reasonable adjustment. With older properties located in historic towns or cities, the lack of onsite parking or appropriately inclusive access route to the buildings is the first recognised challenge. Furthermore, main entrances to these properties may be adaptable for access; however, internal circulation may be difficult. This is more pronounced where there is a need to traverse buildings vertically, and the complexity of retro fitting lifts from a technical as well as cost perspective. It is noted that the provision of facilities for those with disability such as toilets can be achievable where there are existing sanitary rooms that can be altered or adapted. Retro fitting of inclusive measures is further complicated if the buildings are protected by listing or located within a conservation area. In these instances, there is a need to consider honesty in the approach to providing access alteration, which may contradict with the sympathy required in the design to resonate with the historical or cultural importance of these buildings.

Notes

1   England home completions 1949–2021 | Statista (openathens.net)
2   Choose a school for your child | Disability charity Scope UK

References

Scope. (n.d). *Choosing a school for your disabled child.* [online] Available at: https://www.scope.org.uk/advice-and-support/right-school-for-my-disabled-child-specialist-or-mainstream/ [Accessed 14 January 2024].
Statistica. (2024). *Number of housing units completed in England from 1949 to 2022, by sector.* [online] Available at: https://www-statista-com.eu1.proxy.openathens.net/statistics/613564/housing-completions-england-tenure/ [Accessed 14 January 2024].

Chapter 6

Access Audits

Introduction

Chapter 4 detailed the legal requirements as well as the standards that are relevant with the provision of access to buildings, goods, services and employment. Furthermore, Chapter 5 has identified the diverse nature of commercial property and that there is a need to recognise that implementing reasonable adjustment is applicable to most existing buildings. As a consequence, there will always be a need to undertake some degree of compromise. Standards and guidance go above legal minimum requirements, and this has been extensively demonstrated with the summary of British Standard BS8300:2018 in Chapter 4. There should be some acceptance that these are an ideology that can only be fulfilled with the 'blank canvas' of a new build.

Therefore, to establish what should be done to engender access to existing buildings, is it paramount to establish what can be done within the constraints of those building features or characteristics already present. It is necessary to disregard what 'could' be done with what should be done, and this goes somewhat against the advice of professionals or consultants who are often tasked with giving their clients options and allowing them to choose accordingly. Therefore, the 'should' relates to what should be done to comply with best practice to achieve the best possible user-access outcome. To do this, it is necessary to set a benchmark criterion to allow a comparative study to be made of the existing building features and access provision.

This simple process is known as an access audit, and although there may appear to be a complexity with the contents or framework of an access audit, the benchmark criteria is either met or not. There should be a binary 'one or zero' approach to establishing compliance against the benchmark data. The facet of building access either complies or not; it is black and white. An objective, not subjective, approach should crystalise the findings; however, it is always necessary to contextualise the objective findings against the building age, type and a range of other considerations. This is where the auditor displays their professional creditability, and if the result of the audit is a figure representing the percentage compliance of a building to the best practice requirements, it is the subsequent discussion on why compliance has not been met that will shape the necessary remediation required. To ensure that both the client and consultant (auditor) have clarity in what is tasked in an access audit, it is paramount to adopt the relevant necessary professional etiquette.

DOI: 10.1201/9781003239901-6

Client Instruction

The principal requirement of a client instruction is to formalise the client's request by instructing the surveyor to undertake an access audit according to a defined scope of services.

The nature of an access audit is such that this should be universal in the benchmark criterion. It will utilise a methodical and systematic inspection or analysis of an asset, so it is feasible to use a generic format for client instructions that is transferable between different properties. The auditor should make the client aware that the access audit does not replace the obligation for the building to comply with statutory procedures and will not amount to or replace the relevant requirement for obtaining statutory compliance reports. The nature of the audit concerns accessibility, and it should be clear that this will not comment on fire safety or health and safety concerns unless this interacts with the provision of accessibility. Outside of the instruction and if a surveyor observes significant health and safety breaches or fire risks that potentially pose a life safety risk, they should raise this matter with the building owner or operator as part of their civic duty of care as expected by their professional body.

Referring back to the audit and to ensure that clients receive the most appropriate advice, auditors should engage in an initial discussion to develop an understanding of their clients' requirements or even philosophy on accessibility. While businesses are increasingly keen to showcase their approach or corporate policies on diversity and inclusivity, some may simply attempt to pay 'lip service' to this idea. So long as there is no mandatory requirement to seek access solutions and promote an inclusive environment, this will always be instructed, in theory, by those wanting to make a difference. There may also be those who seek an access audit to assess the compliance with the legal minimum, and accordingly, the audit criteria can be adjusted. However, in the best interest of promoting inclusivity the initial client discussion could seek to explain the difference between legal compliance and optimal inclusivity with a need to temper this the notion of 'reasonable adjustment'.

It is beneficial for the auditor to begin by requesting information about the property from the client. This should include confirmation on ownership and any lease arrangements, as well as any planned or proposed works such as refurbishment that may affect the access strategy. Auditors should also establish the tenure, type, nature, size and age of the property. There is also a requirement to consider any specific characteristics such as historical value or listed status and whether this is located within a conservation area. An understanding of the characteristics and complexity of the premises will allow the auditor to estimate the time required to undertake the onsite survey and report. An appropriate fee can then be calculated, reflecting the timescales involved.

The access audit does not usually analyse the service provided; it focuses on the premises where the service is provided. However, specific use classes will have specific access requirements that need to be audited. For example, accessible bedrooms and the associated audit criteria apply to the leisure / hotel sector but are not relevant to offices. While there are black and white examples of accessible features that are relative to the destination use of service providers, others require more thought. While not all retail premises have customer toilets or are obliged to provide this service, they will have staff toilets and, accordingly, the absence of an accessible facility will potentially discriminate against employing those with physical or mental impairments.

A way to alleviate this issue would be to use a proforma or checklist to establish the building type as well as the service provided to clarify the audit criteria.

Clients or users obtaining accessibility advice for their buildings and the delivery of goods, services or the provision of employment should receive the following benefits:

- a statement of the compliance of the building or premises against the clear audit criteria for legal and/or best practice access;
- a statement of potential improvement in compliance with appropriate recommendations; and
- a contextual discussion detailing the reasons for non-compliance and the factors affecting it, as well as the options for adopting measures to provide a 'reasonable adjustment'.

It should be recognised that while the audit process is binary regarding compliance, it is the contextualised assessment and the evidence-based recommendations of the auditor that with give some 'direction' to the findings. On the assumption that clients or building owners and service providers have actively engaged in seeking out an access audit, they are more likely to be receptive or proactive in seeking solutions to promote inclusivity.

While it is noted that there are clients or service providers who are only seeking to verify legal compliance, by adopting this practice with best practice compliance, there is an open transparency to the compete benchmarking process. Accordingly, the client or service provider should be informed of the associated risk of not providing access to those with a physical or mental impairment in accordance with the Equality Act 2010.

The client instruction should make explicit reference to any limitations of the audit/ survey in the sense that it is necessary to gain access to all areas relevant to the building use, operation or service provided. In the event that it is not possible to access specific and relevant area, the instruction should state that there will be a subsequent limitation to the findings or recommendations. As the client instruction is a contractual appointment document, the following items should be included, irrespective of the property type or sector:

- names and addresses of the contracting parties ('consultant/auditor' and 'client');
- date of instruction;
- scope of works: precisely what services will and will not be provided;
- items to be inspected or excluded;
- any limitations;
- programme or timescales for the inspection and delivery of the audit report;
- format of the report (e.g., hard copy, PDF, Excel, interactive database or a combination of these formats), with an example;
- specific access requirements/health and safety;
- specialist access requirements and restrictions;
- fees and variations to the contract, including the provision for extra works; (additional site visits or supplementary meetings, etc.);
- provision for payment terms and late payment;
- disbursements (specialist equipment, travel, accommodation, etc.);
- costs associated with subscription to any interactive databases or online access;
- any standard terms of business;

- prequalifying requirements such as non-disclosure/confidentiality agreements, work permits or specific training certificates/courses;
- signatures of all relevant parties to the contract; and
- level of liability and any other cover and limitations.

The following items may also be included:

- evidence of professional indemnity insurance (PII);
- dispute resolution;
- personal guarantees for payment;
- advance payments; and
- reliance letters or agreements.

Scope of Services and Checklists

One of the most important features of the client instruction is the scope of services. This may be the most bespoke part of the instruction and auditors should tailor it to the needs of the client but also the building type or nature of the service provider.

The key requirement of the scope of services is to expressly clarify items that will and will not be included in the access audit. Accordingly, there will be a need to establish if the audit is to assess the access provision in accordance with the legal minimum requirements or best practice standards. As previously indicated, a combined audit criteria are the most appropriate to firstly comment on whether the legal minimum is achieved, with a view to assessing then advising on the optimal way to achieve the best possible levels of inclusivity. Auditors will need to consider including the following listed accessible features and subcomponents. However, it is paramount to tailor this list to the specific building sector or nature of the service provision:

- Parking:
    - number of dedicated accessible spaces;
    - types of parking bays;
    - off street parking;
    - parking bay dimensions; and
    - ticket machines and barriers.

- Access routes:
    - widths;
    - gradients;
    - hazards;
    - signage
    - surface materials/construction; and
    - barriers or restrictions.

- External vertical circulation.
    - steps;
    - stairs;

- ramps;
- escalators; and
- lifts.

- Building entrance

  - Identification;
  - doors; and
  - reception desk and waiting areas.

- Horizontal circulation

  - Corridor widths;
  - passing places;
  - floor, wall and ceiling finishes;
  - lighting; and
  - doors and escape routes.

- Vertical circulation

  - Steps and stairs;
  - ramps and slopes;
  - handrails; and
  - lifts.

- Signage and communication

  - Audible, visual and tactile signage;
  - hearing loops/communication systems; and
  - alert/alarms.

- Facilities and sanitary rooms

  - Seating & waiting areas;
  - storage facilities;
  - ATMs and cash operated devices;
  - touch screens;
  - windows and their operation;
  - public telephones;
  - outlets, sockets and switches;
  - assistance dog facilities;
  - audience and spectator facilities:
  - shower rooms and bathrooms;
  - changing and shower areas;
  - accessible baby changing facilities;
  - toilet accommodation;
  - changing places toilets; and
  - individual rooms (kitchen areas, accessible bedrooms and quiet spaces).

It is important to consider that each and every item included in the audit criteria will contain specific requirements, dimensions or data. Accordingly, while there is a need to consider the dimensions required to afford access, there is also a need to address colour, contrast, texture and lighting, etc. Therefore, it is necessary to adopt a multidimensional approach to the audit, and this should be evident in the audit checklist. To produce an accurate assessment of the relative accessibility of a building, it is necessary to apply a forensic approach to the data collection; it should include a significant amount of detailed criteria

The scope of services should also explicitly detail items in the access audit and report that will not be included, such as the operations of the service provider or where the audit concerns the entrance and common parts of an office building; it may not include individual tenant demises, which may be covered by separate instructions.

The Survey/Audit and Safety Procedures

The client instruction should detail the requirement for safe working access, and as the audit is concerned with access, there should be very few instances where this should present a health and safety risk to the auditor. However, the auditor should make pre-visit access requests to the owner or service provider to see if there are any operational health and safety concerns.

In the client instruction, the key dates required for the issuing of the report should be confirmed. Additional dates such as those for visits, submission of the report, and payment of invoices may also be included when detailing timescales in the contract instruction.

In the event that timescales are imposed by the building owner or service provider, it is prudent for auditors to assess the timescale and inform the client about whether it is possible to achieve the objectives prior to confirming the instruction. To manage the process and the expectation of stakeholders in respect of the timeline, that auditor should suggest and agree to an appropriate and reasonable time to discharge the scope of services. The agreed timescale should be subject to gaining the necessary access to the premises to inspect all relevant accessibility features. If safe access cannot be granted or areas of the property are inaccessible, the client instruction should allow for the provision of additional time and cost, if applicable.

So long as auditors and surveyors are engaged in the delivery of advice and are professionally liable for their advice, it is necessary to consider the subsequent payment for this work. The client instruction should ensure clarity concerning professional fees, irrespective of whether payment is a fixed fee or calculated on an hourly or day rate. It should state whether tax is applicable and whether disbursements are included or excluded. Where disbursements such as travel to the site or subsistence and overnight accommodation are necessary, the value of these should be defined and agreed in the client instruction. Concerning extra work, there should be sufficient detail in the client instruction to address the basis for the calculation of additional fees (hourly or day rates), as previously detailed.

Both the building owner or service provider and the auditor should acknowledge the contents of the client instruction by entering into a formal agreement prior to commencement of the audit.

Audit Preparation

Before undertaking the access audit, the auditor should obtain or agree to a formal instruction from the client to proceed. This agreement should be in accordance with the client instruction as previously detailed and should be clear and unambiguous regarding the agreed scope of services.

Those undertaking access audits should have the relevant training and experience to be deemed competent to do this work. While it may be argued that the audit process, which comprises a series of observations and checks against a listed criteria, requires relatively little training, there is a need to look beyond this idea. Because the first part of the audit assesses the current situation against a fixed criterion, the second part requires the auditor to make evidence-based recommendations. Accordingly, there is a need to discuss and contextualise the recommendations, bearing in mind the building type, existing features, nature of service provision and the notion of reasonable adjustment. Auditors should therefore have the relevant experience for the specific property types and sectors. They should be sufficiently skilled to inspect and report on all aspects of the access requirement; this should include appropriate knowledge of building technology when considering remedial alteration.

For large commercial properties or portfolios of buildings or institutional service providers with multiple outlets, it is likely that more than one auditor and professionals from other disciplines may work together. Therefore, it is important to ensure that the group or team members are briefed on, and are familiar with, their relevant sections of the scope of services as well as the audit format. To ensure consistency in the application of the audit criteria, and more importantly, any advice regarding recommendations, it may be appropriate to appoint an individual person or central coordinator to brief the team members, allocate tasks and collate the findings or reports. Clear lines of communication should be established between members of the team from the outset to prevent errors and engender uniformity in data gathering.

All individuals involved in the auditing of multiple properties or working for service providers with multiple outlets should be briefed on the following:

- site locations;
- any specific access requirements;
- purpose of the audits;
- names and contact details of the audit team;
- estimated time required for the visits, reports; and
- information concerning any known site risks or hazards.

Ideally, auditors should endeavour to obtain (where available) copies of the as-built floor plans so that these can be taken to the site. These can be used to annotate the information to include key measurements or dimensions, and observations. It is however recognised that there is often an absence of floor plans, and while auditors will essentially work off a checklist, it is prudent to sketch onsite some of the key dimensions and features. This sketch, along with photographic records, will aid significantly in the post visit reporting process.

Equipment

Before the site visit and undertaking the audit, auditors should obtain the equipment required to undertake this and record data. This may include:

- a notepad, pen, pencil or tablet/phone with appropriate apps;
- a camera or phone/tablet with sufficient capability to produce quality photographs;
- measuring devices such as laser tapes and tape measures for taking check dimensions; and
- devices to record lighting levels as well as colour and contrast.

Health and Safety

Unlike building surveys or inspections that might involve accessing roofs or exposed areas as well as concealed spaces, access audits should be relatively low risk. However, before undertaking them, auditors should undertake a risk assessment based upon information that can be gathered about the property. If it is not possible to visit the site before the survey, a desk survey can be undertaken with the use of online satellite and street-view information gleaned from internet searches, although images may be date limited. The reality is that health and safety risks concerned with auditing parking and access routes may be presented by onsite vehicles. Access routes may present risks of slips and trip and falls, but in reality, it is anticipated that the buildings and service providers are already in use. Accordingly, the types of risks mentioned should have been considered and addressed; however, when undertaking access audits of places of work, these might pose service specific risks. For specific industrial or manufacturing sites, there are likely to be highly controlled health and safety procedures in place; auditors may therefore be obliged to comply with onsite health and safety procedures.

The Inspection

While there is a bespoke nature to many buildings and nuances to service providers, the principle of undertaking an access audit is universal: to undertake a methodical examination of the key access requirements against a standardised checklist. That said, auditors should have sufficient experience and knowledge of building components and materials to form post-audit advice on how to facilitate or improve access. For example, recommendations to widen door openings to walls will need to consider if these are likely to be load bearing as such advice could be complex and costly to implement. Likewise, the removal of obstacles such as columns or pillars could result in compromising building stability.

As part of the audit procedure, it will be necessary to check if the building is listed or located within a conservation area as this status needs to be considered when preparing advice. Chapter 8 specifically discusses access to the Historic Built Environment (HBE); it should be noted that while there are restrictions placed on making alterations to a protected building, there are also quite progressive thinking and proactive application of reasonable adjustment. When reporting on this idea and discussing the options to provide step-free access, the auditor is required to think outside of the binary box that is the audit checklist to look for evidenced-based, proactive solutions.

Unlike the time pressures afforded to surveyors and auditors working to inspect buildings for acquisition, where pressure is applied by the vendor or purchaser as part of the process to speed up closure of the transaction, access audits should be easier to time manage. Upfront research and preparation of the client instruction should give the auditor sufficient knowledge on the time allocated for the inspection. Clearly, something like a modern, purpose-built, single-storey retail unit should be quicker to audit than a complex multi-storey leisure sector building, such as a museum, art gallery or hotel. Accordingly, it is necessary for the auditor to agree to a fee with the building owner or service provider to reflect the time needed to complete the audit. In theory, and provided there are no access restrictions placed on the auditor to complete the checklist, there should be no reason why the relevant data is not collected during the inspection.

Such is the very nature of an audit that it is necessary to systematically inspect the property, completing the access audit checklist, taking notes and recording photographic evidence. Notes may be recorded in written form, dictated or using tech devices such as smartphones or tablets with cloud-based inspection/reporting software. The audit criteria can be addressed by completing check boxes on a proforma, including numerical information such as dimensions. Additional notes should also be used to contextualise the findings recorded onsite and should be used to develop the survey report. Auditors should retain a copy of their original, unedited survey notes on file as it may be necessary to return to these for reference and also as a requirement of the professional indemnity insurance providers.

Electronic data capture software may be used for access audits, and the nature of using a standardised list lends itself well to being developed for electronic capture and may prove particularly appropriate for larger, multi-building programs. Here, multiple auditors or audit programs spanning several company office locations, or service providers with multiple outlets, can benefit from using tools that increase consistency.

Rather than a traditional pen-and-paper approach, electronic data capture involves apps on mobile or tablet devices. With base data captured onsite and the tools available to manipulate that information into dashboards, there is also the possibility to integrate qualitative data in the form of comments or observations in reports. IT systems can provide auditors with the ability to provide consistent data and outputs across portfolios. Collecting information in this way can ensure multiple auditors are working to the same element lists in real time, using the same terminology to ensure consistency of the data and report outputs.

Auditors should be aware that software and the use of electronic devices are tools to assist in the collection of data, and accordingly, professional opinion will always be necessary to deliver evidence-based advice and recommendations.

The physical survey associated with the audit has the prime purpose of establishing the current access provision; however, there is a greater need to multitask this process and look towards any recommendations for improved inclusivity. Therefore, it is paramount to identify the material types of the construction elements to gauge the appropriateness of these and if they are fit for purpose. As previously discussed, and in the context of recommending a reasonable adjustment, there may be a need to consider means to avoid a barrier or seek an alternative access. This, along with possible removal or alteration of a barrier, may have significant implications on the fabric and structure of a building.

Each access element should be considered individually against the criteria set for the audit and considering the logical flow and interconnectivity between the different elements. To get from onsite or street parking to access goods or services to the upper floors of a building requires the navigation of a number of potential barriers. Each part of this physical journey is connected, and it is necessary to ensure that this connection is evident in the subsequent commentary accompanying the audit.

Element	Common Audit Characteristics
• Parking	• Base dimensions
• Access route	• Element specific features
• External vertical circulation	• Colour
• Main entrance	• Contrast
• Horizonal circulation	• Texture
• Vertical circulation	• Light
• Signage & communications	• Communication and alert
• Facilities and services	

Note: Within every element, there are a number of critical sub-elements which have been previously detailed in this chapter; therefore, generating an audit checklist that covers the range of common audit characteristics will no doubt generate a lot of data points. A high quantity of data will add significantly to the overall quality of audit report and enhance the credibility of the evidence-based findings. As with many surveying applications, it may be possible to delivery general high-level advice in the form of an executive summary. However, collecting and recording component level data means it is possible to drill down to the necessary detail if the findings are the subject of cross examination.

The Report

After undertaking the site inspection and having collected the necessary data, the auditor should collate all of the information into an easily readable document. This should effectively be divided into two separate parts, with one section devoted to a representation of the access audit checklist. The other section is the more formal report that discusses the basis of the instruction, building details, and importantly, the notion of reasonable adjustment. While the audit is binary and either states compliance or not, the written report discusses the constraints and limitations of the existing building features, as well as recommendations to address non-compliance. While the written representation of the checklist clearly identifies the level of compliance, there should also be space within this document to indicate if compliance can be met with reasonable adjustment. To an extent, the checklist in itself can be a stand-alone report on accessibility that can accompany an executive summary. However, because of the nuances that exist with individual buildings or service providers, a rounded, evidence-based discussion is better suited to a full written report with the checklist in the appendix.

Accessibility Schedule (Checklist)

The most appropriate way to illustrate the accessibility of building elements, their compliance, and whether compliance can be achieved with reasonable adjustment is with an accessibility schedule. As there is a need to illustrate percentage compliance and areas of concern, it is appropriate to use a spreadsheet. This makes it possible to insert, alter or omit compliance issues relative to the building or service provider.

Reference	Element/ Location (of the accessible feature)	Description (of the prescriptive requirement in the legislation of guidance)	Observations (data recorded)	Compliance (1 or 0)	Potential Compliance with reasonable adjustment (1 or 0)	Photo Ref

It is important to ensure that the base reference data for the audit is included in the title of the accessibility schedule. The reference can relate explicitly to a paragraph or clause numbers in the relevant legislation or guidance, but this may complicate the document as there is expected to be potentially over 100 data points. The reference is typically used where there is a need to discuss specific issues in the report or for cross-examination purposes. Logically, the reference numbers should be incremental, with each accessibility element of the schedule having a different number and corresponding sub-reference.

The base access elements should be included for all buildings, and in essence to ensure consistency, every accessible schedule should have the same base format. Clearly, there may be differences in the overall audit process between different sectors or providers with sectors, as well as situation-specific considerations. There is a need to detail or describe the location of the accessible feature, and this may be through description (such as *Main Entrance*) or through more specific geographic reference such as *North Entrance*.

The accessibility schedule should include a description of the prescriptive requirements aligning with the base legislation or guidance. This, along with the locational description, will begin to create an understanding of where and what the access requirement is. The observations recorded during the audit should be detailed, and this may be through descriptive text and/or details of dimensions and actual data collected. This will generate the evidence-based conclusion on whether the existing situation is compliant.

While the process of the audit is the collection of data and comparing this against the base requirements to create a binary conclusion on compliance, assessing whether compliance can be achieved through 'reasonable adjustment' is where the worth of the auditor is realised. Auditors with good technical knowledge, experience or from professional surveying backgrounds should be able to make measured, evidence-based recommendations. It is these recommendations that may be forensically cross-examined in the event of a complaint or dispute. While there may be no precise right or wrong recommendation, the auditor should be prepared and equipped with evidence to back up their advice on achieving compliance. It is necessary to refer to some of the previous guidance on adopting measures to deliver a reasonable adjustment or facilitate access. In essence the recommendations should seek to:

- Remove or alter a physical feature or provide a reasonable means of avoiding a physical feature that puts a disabled person at a substantial disadvantage when compared to a non-disabled person. This should be done on the basis of:
  - the overall effectiveness of this step;
  - the practicality of implementation;
  - financial, other costs and disruption on implementation;
  - employers available financial or other resource;
  - availability of financial or other assistance to make adjustment; and
  - the type and size of employer.

Added to the nuance of undertaking reasonable adjustment to facilitate access, it will be necessary to consider potentially overriding legislation or principles related to fire safety, health and safety, as well as listed building legislation but to name a few.

The notion of compliance being binary means that it is necessary to use the number '1' for access provision being compliant and a '0' for non-compliance. By adding up all of the numbers and dividing this total between the maximum number equating to full compliance can be represented as a percentage. With all spreadsheets and the automatic function to collectively add numbers and create percentage compliances, it is necessary to cross-check the calculation. This cross-check is important where accessibility elements or sub-elements are not applicable and should be left out of the calculation or given neutral status (neither compliant or non-compliant). In simple terms, an office does not have bedrooms, so there is no need for these to be accessible because they do not exist. Therefore, this item or sub-item should be not applicable. A similar process may be evident with a simple retail unit that does not have publicly available toilets and therefore should not be penalised for the absence of an accessible cubicle. In contrast, a restaurant with sanitary rooms should ensure that there are accessible toilets and they are measured against this benchmark to confirm the relative percentage compliance.

Below is a fictional example accessibility schedule that only addresses three individual access items making up the 'Access Route' category. This identifies the reference, element location, requirement and relevant compliance. There is an additional column for a photographic reference, and this can prove particularly valuable if someone else other than the auditor reads the report. It adds significantly to the ability to contextualise the accessibility situation, as well as giving an understanding of the evidence used to make the relevant recommendations.

Item 1.1.1 discusses a secondary access route, where there is no pavement, in this instance, a reasonable adjustment to prevent pedestrians and vehicles from colliding would be to install and mark off an access route with bollards. This is a subjective attempt to address 'reasonable adjustment', and the discussion behind this idea can be presented in the accompanying written report.

Disabled Access Provision – Reference (Legal Regulation or Best Practice Guidance)

Ref	Element/ Location	Requirement	Observation	Compliance	Potential Compliance	Photo Ref
1	**Access Routes**					
1.1	Main external access route from parking	A separate pedestrian route minimising the danger of walking into a vehicular access route	Separate pavement running parallel with the road, raised above the road with fixed kerbstones	1	1	2,3-5
1.1.1	Secondary access route to rear of building	A separate pedestrian route minimising the danger of walking into a vehicular access route	Access from parking spaces is not separated from the access road	0	1	6 & 7
1.2	Access Route – General	Min. 150 mm (h) upstands along access routes	No raised upstand to pavement which is under local authority ownership. Placing a raised upstand could introduce a trip hazard	N/A	N/A	

Current Percentage Compliance   50.00%

Potential Percentage Compliance with Reasonable Adjustment   100.00%

Example of a Partial Accessibility Schedule

Item 1.2 in the example accessibility schedule identifies the requirement in the access guidance for 150 mm upstands along the access route. While this requirement could be implemented for the access route on private land or perhaps on a bespoke access path leading to the building entrance, within the domain of the local authority, such access provision is outside the scope of the audit. Irrespective of this, the placement of a raise edge to an access route does introduce the possibility of a trip hazard. This no doubt would be the case if such design were implemented to all pavements. The need to visually highlight the edging and add sufficient lighting for the entirety of the pavement may reduce the risk of the trip hazard but not remove it. Therefore, the auditor has sufficient evidence and knowledge to form an opinion. Such instances of assessing whether a reasonable adjustment can be adopted are common, and if the auditor is not comfortable proposing design solutions, they can default to reporting on only the level of compliance. However, it will be necessary to discuss the reasons for non-compliance in the written report, referring back to the relevant audit benchmarks for compliance.

Access Audit Report and Executive Summary

It may be argued that the percentage compliance of a building for the provision of goods, services, and employment is the most important fact-based finding. There is a sense that irrespective of how bad it may be, there is always likely to be the opportunity to improve it. Therefore, to introduce and discuss the findings of the Accessibility Schedule appended to the report, an executive summary may be used to outline key information about the property and principal observations. This summary is done by means of delivering the headline findings of the audit and any key non compliances. The nature of an executive summary is such that it can be added to the Accessibility Schedule to create a stand-alone document, but more often, it is used as the lead into the main detailed report that discusses the findings in full.

The executive summary may include:

- summary of scope of services and the base legislation or guidance used for the audit criteria;
- the nature of the property, its construction age and design;
- the adequacy of the key accessible requirements and percentages of compliance;
- principle areas of concern;
- a summary of key recommendations; and
- the period the access audit will last for and at what date it should be revised.

It should be noted that the access audit is only valid as long as there are not significant changes or alterations to components that constitute the base accessibility features or elements. One example may be the renovation of a façade and new shop frontage, which might alter the entrance door, although if this is subject to building regulations approval, the subsequent dimensions should be sufficient or even an improvement. The division of internal space may introduce difficulties with horizontal circulation and opening up extra upper floor space could be significant in respect of vertical circulation. Most of these changes are likely to occur with the change in owner, occupier or service provider;

accordingly, the entity named on the access audit may change and this would require a new audit.

As an optional addition to the Accessibility Schedule, and if instructed by the client to write a full report, auditors should commence with an introduction, followed by the main body or discussion, with a short conclusion used to sum up the findings. The introduction should frame or contextualise the survey by referring to the date and basis of the client instruction, and it should touch upon any specific limitations that may have influenced the survey procedure or outcome.

The report should also include a brief description of the property and its location before summarising the principal access observations element by element. It can be used to detail the issues over non-compliance linked to the specific building elements, which may relate specifically to the construction technology associated with the building age. The report can discuss the notion of reasonable adjustment in analysing the option to remove or alter barriers or in specific instances seek alternative access solutions. It is the first opportunity for the surveyor to express, in writing, the consequences and reasons for non-compliance. This report will enable the building owner, occupier or service provider to understand the current situation based upon the evidence presented to assess the options to achieve or improve accessibility.

The report should have a clear conclusion that is objective and reflects the outcome of the site survey. This is another opportunity to reiterate the principal findings, concerns and recommendations relating to the main body of the report and accessibility schedule.

An access audit is the starting point in the realisation of the actual level of compliance regarding an inclusive environment. Looking deeper into to overall reasons for creating an inclusive environment, published research has established both public and commercial attitudes to disability in the Built Environment, and this will be discussed fully in Chapter 7.

Public and Commercial Attitudes to Disability: Accessing Goods, Services and Employment

Introduction

It has been established that there is an appropriate and clear definition of disability in the UK. Although there are nuances in some of the subsequent wording contained within the legislation, for example, the very nature of the word 'reasonable' used in the context of making adjustment for access provision does appear to lack clarity. Providing access for all and full integration of disability in accessing goods, services and employment is the epitome of an inclusive environment. Surely, full integration is a win: a win for society and those delivering goods, services or employment. However, and because the provision of help to those with disability has proven to be so polarised over centuries, it is necessary to establish both public and commercial attitudes to providing an inclusive environment. Prior to establishing such attitudes, it is necessary to view disability through the lens of those without an physical or mental impairment and how society has come to view disability.

Background and Showcasing Disability

Considering the historic portrayal of disability, it can be suggested that irrespective of polarised efforts to help or hinder those with a physical or mental impairment, disability was rarely portrayed in a good light. The use of derogatory terms recorded in Medieval England allied with the notion that disability was seen as a punishment from God helped create the negative narrative. As previously discussed, the depiction of Richard III as a frail and weak king was perhaps based more so upon his medical condition of scoliosis, this despite the disputed notion he was actually quite proficient as a medieval monarch.

Seen as a weakness, evil or a drain on society, it took a cataclysmic event to begin to change public perception on disability. The devastation of a world war undertaken on the fields of Flanders was a rude awakening, with hundreds of thousands of much-loved British boys returning from battle with life-changing injuries. There is little to be celebrated in war; however, one overwhelming positive was a change in attitudes towards disability. The need to build homes for heroes, societies for benevolence and the good will of humanity to provide employment for those with physical and mental impairments were a starting point to the integration of disability. Relative advances in medical treatment or intervention helped to provide practical solutions to disability. However, despite this, there is some suggestion or feeling that society viewed those disabled by war differently that others with congenital physical or mental impairments.

The turbulent economic interwar years soon gave way to global conflict again, this

DOI: 10.1201/9781003239901-7

time far-wider-reaching with even more deadly consequences. Once again, there was focus on wounded soldiers, but this time, it was with an attempt to disprove or tear down the barriers of disability in the participation of sporting excellence.

The Paralympic Games were formed in 1948 to involve wounded veterans in the London Olympics. Originally named 'the Stoke Mandeville Games', the Games changed to a format in 1960 that we now know as the 'Paralympic Games'.[1]

Fast forward 16 Olympic cycles and 64 years to the 2012 London Paralympic Games, a landmark event for disability sport. The event has been established as recordbreaking in ticket sales, commercial investment and driving change in attitudes to disability.

Despite perceived changes in attitudes by essentially those without a disability, post-2012 emerged contradicting opinions regarding the legacy of the Games and tangible improvements for those with a physical or mental impairment. With the passing of time and economic turbulence resulting in austerity, disability appears more than ever to polarise opinion. Published articles questioning public attitudes to disability discussed the notion that society views those with a disability as either superhumans or scroungers. Noted also is the emergence of a perception gap between the disabled and non-disabled regarding discrimination. Essentially those experiencing disability perceive there to be more discrimination against them than the perception of those without a disability, who believe there are lower levels of discrimination.

Commercially, goods and service providers have had to deal with the rise of online shopping, the COVID-19 pandemic and spiralling overheads. Added to this are the pressures of increased energy costs and a need to consider retrofitting commercial property to enhance energy efficiency with a long-term aim to achieve carbon net zero. All of these factors appear to have placed disability discrimination and the access to commercial properties, goods, services and employment away from the forefront of both the general public and those owning, operating and advising on commercial property.

In light of the apparent stagnation in both the attitudes and actions associated with delivering an inclusive environment, research was commissioned by The University College of Estate Management through the Harold Samuel Research Prize into *Public and Commercial Attitudes to Disability in the Built Environment*.[2] The findings of the research undertaken during the COVID-19 pandemic were published in 2020, identifying a number of key observations that will be discussed in this chapter.

Prior to examining the research, it is necessary to recap on a number of key issues already detailed in this book. Accordingly, it has been established in Chapter 4 the base legislation concerning the prevention of discrimination of those with a physical or mental impairment in the context of access to goods, services and employment. It is also necessary to begin to contextualise the legal provision and its adoption into the different commercial property sectors, which were detailed in Chapter 5.

Commercial Property, Real Estate Investment and Disabled Access to Goods, Services and Employment

As discussed in Chapter 5, a wide range of commercial properties can be broadly categorised into the following sectors:

- Residential.
- Retail.

- Offices.
- Industrial.
- Leisure.
- Public sector/infrastructure.

While there are fundamental differences between some of the commercial properties within each sector, there are also nuances between the properties in the same sectors with respect to the function they perform and the construction characteristics. However, all commercial properties have one defining feature, which is that they are operated as a business. As a consequence, the 'bricks and mortar' as opposed to the service delivered are an attractive proposition for real estate investment. Having a fully leased commercial properties with tenants' rent providing an income stream is not a new proposition, and as an investment vehicle, real estate is seen as an attractive, safe mode of investment.

Irrespective of the service provision, buildings are designed or conceived for a function, and accordingly, there is a planning system that permits certain types of development into the various use classes. Regulations are in place to ensure conformity with legal requirements, and accordingly, there are a set of approved documents known as the Building Regulations. These are generally divided into the regulations covering housing and those covering *buildings other than dwellings*. In most cases, these regulations apply to commercial properties, and as previously discussed in Chapters 4 and 6, the approved document governing disabled access is known as *Part M*. While there are regulations affecting the new build or renovation of commercial property, there is an increasing awareness in the construction industry and real estate investment market to improve accessibility. Despite this, it may be commented that minimal tangible improvements appear evident in society, with accessibility seen as a box-checking exercise.

As reported by the Construction Industry Council (CIC) on 25 April 2017,[3] there was a recognition by the UK government in its report; *Building for Equality: Disability and the Built Environment*[4] that accessibility is an issue. In essence, *Much more can be done to make the public realm and public buildings more accessible* (CIC, 2017). Included in the buildings defined as being within 'the public realm' are buildings associated with the public sector, as defined in Chapter 5. The government report highlights the need to update regulations for new buildings, based on a 16-year-old standard and amending the Licensing Act 2003 to force landlords to make their properties more inclusive.

The commercial property sector is so much more than public-realm buildings, but despite this, one general observation is that it is the public sector buildings that are more likely to adopt an access for all approach than private sector commercial property. It is important not to underestimate the purpose or reasons behind commercial property as and investment and the notion that, in most cases, there is a disconnect between the building owner (landlord) and service provider (tenant). There are relatively few owner occupiers, therefore commercial property is leased with the intention of creating a return on investment.

As previously detailed, the UK commercial property sector is divided into six different sub-sectors: Office, Retail, Leisure, Public Sector, Residential and Industrial. Clearly there appears a need for accessibility provision and, in the case of offices, this is normally provided to the common parts with tenants responsible for the area of their lease know as their demise. However, under The Equality Act 2010, the service provider is also obliged not to discriminate on the grounds of disability. Accordingly, while the landlord may provide full accessibility to the principle entrance of a multi-let office building, the tenant will also have to foresee accessibility from within their premises. The building entrance and internal areas are only one part of the property concerned with access. As detailed in Chapter 6, the following comprise the general auditable building elements or features that require consideration in respect of accessibility (Tagg, 2018).[5]

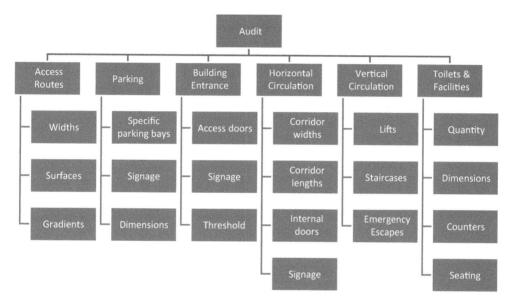

Figure 7.1 **Access audit criteria for commercial property (Tagg, 2018).**

Access to the common areas should be provided by the building owner or landlord (subject to lease conditions), and individual tenants as service providers or employers may be obliged to undertake the necessary reasonable adjustment to their individual demises. Furthermore, and to recap, retail properties are categorised as traditional high street or shopping centres. Although the application of the EA10 should be the same regarding *reasonable adjustment*, shopping centres are in the main multi-let properties with significant common parts that should be treated similar to offices. Individual units or shops within shopping centres should comply with the EA10 in accordance with their function as either employers or service providers. Alternatively, 'traditional' high street retail properties are likely to comprise single tenant occupiers and are often accessed directly from the street. These properties should be treated as a single entity or demise for the service provider regarding accessibility.

Commercial properties within the industrial sector are largely manufacturing or warehouse and logistics. While it may appear that there is little need to inclusivity for these areas, it is important to consider that there is still an obligation placed upon the employer or service provider to facilitate it. Most industrial buildings include an administration or office area, and accordingly, these should seek to include the disabled access provision. The leisure industry and civic buildings, which are publicly accessible, are amongst some of the most high-profile properties in the UK. They and also the services they provide should be fully accessible to those with a physical or mental impairment.

UK real estate investment is estimated to have gross added value (GVA) in excess of £1bn in 2019 or 7% of the national GVA, which is considered a proxy for GDP.[6] The emphasises the importance of real estate as an investment vehicle and is testimony to the numbers as well as prevalence of property through the UK. There has been well documented issues with real estate investment post COVID-19 pandemic, and there is no doubt a challenge from online providers. Consequently, a large increase in online shopping, increased high street costs and low rents has contributed to the perceived *Death of the high street* ... as both a user and investment experience as reported in the *Financial Times.*[7] But perhaps this obituary for the high street appears premature with an encouraging return to experiential shopping following the exit from the various COVID-19 lockdowns.[8] One certainty is that to counter balance a reduction in consumer spending in high street stores has seen online sales increase significantly, and in turn, the demand for logistics space has increased significantly. The implications of economic uncertainty in the retail sector may suggest that there is less available capital to invest in access provision; however, in contradiction, the demise of many traditional high street chain stores has resulted in more independent shops offering bespoke user experience coming into the UK retail sector.[9]

Irrespective of negative economic prosperity, there is still a requirement under the EA10 for service providers to ensure reasonable adjustment is undertaken to provide an inclusive environment. In a US-based report by the company *Accenture* titled *THE ACCESSIBILITY ADVANTAGE: Why Businesses Should Care About Inclusive Design,*[10] the concept of *Sustainable Inclusivity* is discussed. It is suggested that creating commercial inclusivity has tremendous opportunity, and this can be achieved through the employment, engagement, enabling and empowering those with disability. The themes contained in the Accenture report were largely discussed in an article published by the *Telegraph* in 2017. Importantly, the article quotes Farrah Qureshi (Global Diversity Practice):

> *Diversity must be a holistic strategy that considers customers, employees and suppliers. Then it's all about delivering.*

There appears the potential for businesses and service providers to have a commercial edge by strategising inclusivity and surely investing in 'reasonable adjustment' should seek to divert away from legal obligation in favour or commercial opportunity.

Despite the obligations placed on employers and offices delivering professional services, inclusivity is not amongst the top 20 criteria for workers,[11] and there is a sense that employees with a physical or mental impairment are 'invisible'. With an estimated relative average small percentage (7%) of the workforce recognising as having a physical impairment affecting their mobility, it is perhaps plausible that accessibility is not high on employee agendas. There is also a sense that employers act retrospectively concerning disabled members of staff in such that access provision is reactive and not proactive. Despite this, it should be noted that over 52% of disabled people aged 16–64 are employed, and the gap between non-disabled and disabled persons in work has been reducing over the past 6 years.[12] It's likely that most people (knowing or unknowingly) have a colleague who considers themselves disabled. There appears potential commercial gain to engaging with disability from an employer or service providers' perspective.

The notion concerning 'reasonable adjustment' has been previously discussed on numerous occasions, but it needs to be clarified if 'reasonable' is from the perspective of the service provider or the disabled end user. Certainly, in the context of over £1bn GVA generated by real estate investment, the cost of 'reasonable adjustment' likely to be a very minor percentage of this, it is not known how investors or service providers view the potential cost benefit to providing inclusivity.

'Fire Door Approach'

In the context of providing accessibility, Building Regulations (Approved Document Part M) provides the legal minimum standards for inclusivity, although British Standard BS8300 is recognised as going above and beyond the legal minimum. This is evident in the discussion detailed in Chapter 4. It should be noted that while building regulations are not typically enforceable retrospectively, these and British Standard are much easier to accommodate with a new build project or major renovation. All other examples of retrospective implementation of regulations are likely to be a compromise. This perhaps is another way to view 'reasonable adjustment' from the perspective of applicable options for an existing building.

As with most prescriptive requirements, it does not appear to be in the nature of developers or investors in commercial property to go above and beyond legal obligation. There is a sense that with an investment vehicle such as real estate, it is advantageous to do the bare minimum to extract maximum profits. This philosophy is typical to the maintenance and repair of commercial property in a practice known as to 'sweat the asset'. Accordingly, the notion of holding back on building maintenance or seeking to apply only the legal minimums is a race to the bottom.

In the context of providing inclusivity, it appears an alien concept for developers and investors to go beyond the legal minimum, and this is exemplified with something like a fire door. If the legal prescription requires a 30-minute fire door, then it makes no commercial sense to install a 1-hour fire door as this increases cost with no perceived commercial benefit. Achieving minimal compliance while delivering profit probably works for most operations in property investment, and this commercial approach is considered the norm. However, with inclusivity, there is the opportunity to challenge the 'fire door approach' and provide facilities above and beyond the legal minimum. If achieving inclusivity can provide added commercial value, then this is a win-win for both investors and society.

Public and Commercial Attitudes to Disability and the Access of Goods and Services

The notion of challenging the 'fire door' approach and also establishing both public as well as commercial attitudes to disability in the built environment was examined in the report commissioned by the University College of Estate Management.[13] The key findings of the report have formed the basis for this chapter, based on data collected in 2019 to include:

- 'Public Attitudes' Survey.
- 'Commercial Attitudes' Survey.
- Interviews with key individuals representing those with disability and designers.

Public Attitudes Survey

To analyse the credibility of the primary data collected from the 'public attitudes' survey, it was necessary to establish some basic demographic information concerning:

- age;
- UK region of residence; and
- recognition as having a physical or mental impairment.

The research data established that the responses came from the following age categories:

- 18–24 17%.
- 25–45 16%.
- 35–44 24%.
- 45–54 25%.
- 55–64 19%.
- 65+ 9%.

The survey response was compared to the relevant UK national data regarding age demographics of the population at the time of the research (2018) and is shown in (Fig 7.2).

Generally, there appeared a good alignment between the research data and that of the national average data. However, noted were variations between the data sets in the 18–24, 45–54 and over 65 categories. This research focussed on the adult population

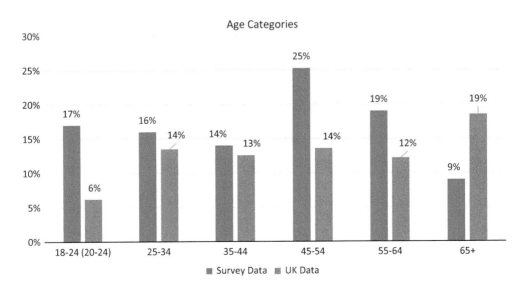

Figure 7.2 **Survey age demographics v published UK age demographics.**

whereas national data concerning age demographics including 18-year-olds coverers ages 15–19. As under 18s were not considered in this research the 'nearest' UK statistics age category is 20–24 which does not take into include those aged 18–20. Therefore, this is likely to account for the difference in the baseline data collected from the survey compared to the published data. There is no known reason why the age category 45–54 was overrepresented or why the over 65s is under-represented. However, there is a link between age and disability, as detailed in Chapter 3, therefore the under representation of the 65+ age category might have skewed the findings.

When considering the relevant age categories of the survey respondents, it is important to note the more generalisation of the age categories published by the Department of Work and Pensions (2018). This focusses on three specific age categories:

- Children (under 18s).
- Working Age (18–64).
- Retirement Age (over 65).

To ascertain the validity of the research as a UK study, the participants region of residence was established. Accordingly, there is a reasonable correlation with the national data although a significant overrepresentation in the Southeast, Southwest, and Yorkshire & Humber. The Northeast is an area of a relative high percentage of disability, but this is noted to be underrepresented in the research data (Fig 7.3).

In essence, it can be concluded that there was majority alignment between the research data and data published by the UK government concerning the percentage of those recognising as having a physical or mental impairment within regional context.

There was a sense that when asked if they have a disability, the elderly are less likely to recognise this compared with being asked if they have a long-term physical or mental impairment affecting ability to undertake normal day-to-day activities. This might

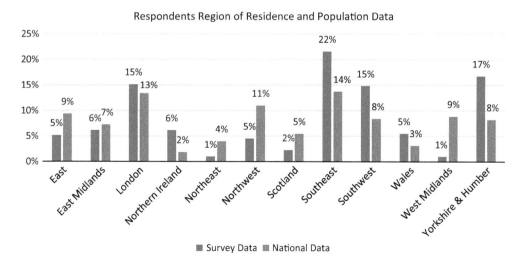

Figure 7.3 Respondents region of residence.

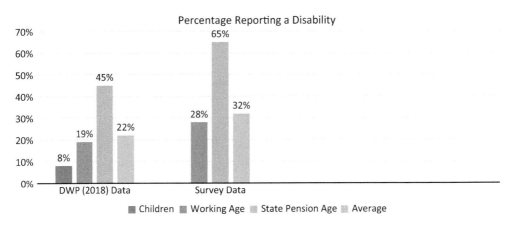

Figure 7.4 Survey response of those recognising as having a physical or mental impairment.

account for the reason that the research established 32% of respondents recognising as having a disability under this definition. The percentage of those reporting a disability (32%) is significantly higher than the published UK government data (DWP, 2018) but more in line with the 30% reported by Dixon *et al* (2018) in the *Disability Perception Gap – Report* published by Scope. This also aligns to the relative high response rate from the over 65s who are more likely to have a long-term physical or mental impairment. Working age was also overrepresented in the terms of those recognising as having a disability. The findings of the research data are illustrated above (Fig 7.4).

The absence of children from the survey has no doubt impacted on the positive skew of the research data, but of significance are the relatively high levels of disability recorded amongst the working age. There is also recognised in Gov UK data an increase in working age recognising mental health as an issue. Accordingly, these two factors may

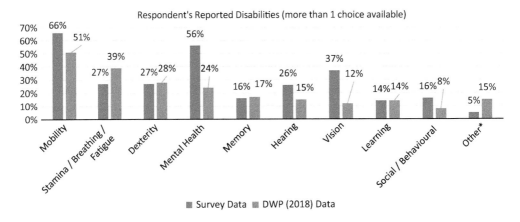

Figure 7.5 Respondents reported disabilities.

be linked and ultimately increase the overall percentage of those recognising as having a disability.

Analysing the responses of those with a physical or mental impairment further, and examining each type individually, per category of disability, has established that those with physical impairments (mobility, stamina / breathing, dexterity as well as hearing impairments) have expressed marginally more disagreement that commercial properties and goods or services are fully accessible. It appears to be less of an issue for those with mental impairments. This difference may be due to those with a physical impairment identifying physical barriers when accessing the built environment.

Looking more closely at the research data concerning the different categories of disability, there is a correlation with national data. However, noted were significantly higher incidences of mobility, mental health, hearing, vision and social / behavioural reported disabilities, as shown above (Fig 7.5).

Accessing Goods and Services

Having established the demographics of those participating in the research and in order to establish attitudes to disability, it was necessary to gauge the frequency of visits to commercial properties (Fig 7.6).

It is evident that the majority of respondents visit commercial properties every day or more than four out of seven days per week.

By undertaking a more forensic analysis of the data relative to the specific age categories, it is evident that age is inversely proportional to the frequency at which respondents visit commercial properties and access goods or services (Fig 7.7).

There exists a notion that younger people are more likely to be more socially and physically mobile when compared to those over 65. Therefore, they are more likely to visit

Figure 7.6 **All respondent's frequency of visits to commercial properties.**

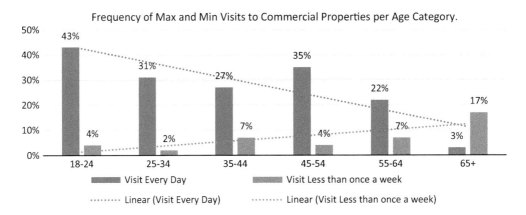

Figure 7.7 **Age categories and the frequency of visits to commercial properties.**

commercial properties, and this is evident in the trend line on the graph. However, this does also illustrate that the 'elderly' visits less often and raises the question of whether this is because they do not want to or simply cannot due to poor access provision.

Further analysis of the frequency at which visits are made to commercial properties in the context of those who recognise as having a physical or mental impairment also produces a trend line suggesting as similar trend of less frequency.

According to the research, it is broadly evident that those who identify as having a physical or mental impairment visit commercial properties less frequently than those who do not have a disability. Further analysis of the data has identified those with mobility and memory health impairments (9% and 13% of respondents respectively) have a higher incidence of visiting commercial properties less than once a week. However,

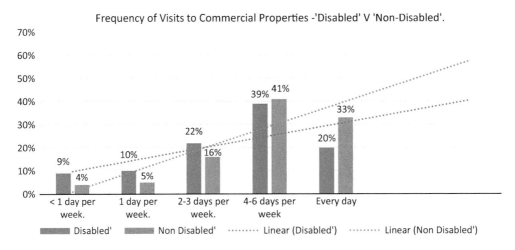

Figure 7.8 Comparison of frequency of visits to commercial properties for those with and without a disability.

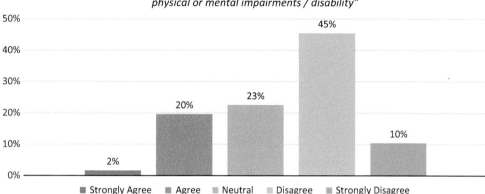

Figure 7.9 Accessibility of commercial property for those with physical or mental health impairments.

respondents in the age category of 65+ and who identify as having a disability visit commercial properties less frequently than any other demographic. With this in mind and the notion that a lack of access provision may affect the frequency of visiting commercial properties, it necessary to examine opinions on the accessibility of goods and services.

When canvassed for an opinion on current levels of accessibility to the built environment, over half (55%) do not agree that this is fully accessible to those with a physical or mental impairment (Fig 7.9).

As indicated in Fig 7.9, and despite the overall majority opinion (55%) that suggesting commercial properties are not fully accessible to those with physical or mental impairments, 23% are 'neutral' leaving only 22% who believe the contrary. Looking deeper into the data per age category, it is evident that with an increase in age, there is a belief that properties are less likely to be accessible (Fig 7.10).

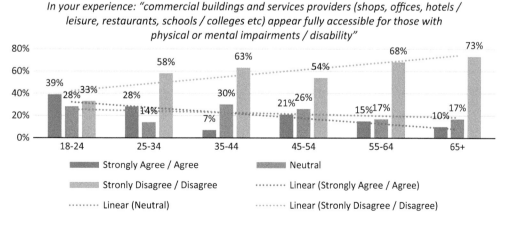

Figure 7.10 Accessibility of commercial buildings and service providers per age category.

When directly put into context of those who recognise as having a physical or mental impairment, there is an overwhelming sense that commercial properties, goods and services are not fully accessible.

A pattern has emerged confirming that those who recognise as having a physical or mental impairment and those over 65:

- access commercial properties, goods and services less often and
- believe that these facilities are less accessible to those with a disability.

Legal Provision and Reasonable Adjustment

Key to understanding public attitudes is to gauge if there is an awareness of the legal provision governing the accessibility of goods and services. According to the research data, **90% of respondents are aware of** the requirement for service providers to make 'reasonable adjustment' to allow for accessibility.

When questioned further on who should be responsible for the funding reasonable adjustments, it was noted that the single clear individual responsibility is that of building owners (73%), as shown in Fig 7.11. However, there is a sense in the data that there is also a responsibility for funding to come from both central and local government as well as the service provider.

Despite some suggestion that funding should come from the taxpayer, the conclusion of data from the public attitudes survey was that building owners should be responsible. The pragmatic approach that those proving the goods and services should be responsible for funding access appears a sensible option. There is little encouragement or appetite for commercial organisations to do this, and in a low tax economy such as the UK, it is hard to imagine funds or subsidies being made available for this. Accordingly, when asked if the current levels of funding for accessibility are sufficient, **the majority of respondents (62%) selected the answer the funding is not sufficient**. Only 8% answered that it is sufficient, and 30% or respondents were 'unsure'. This therefore places an ownness on

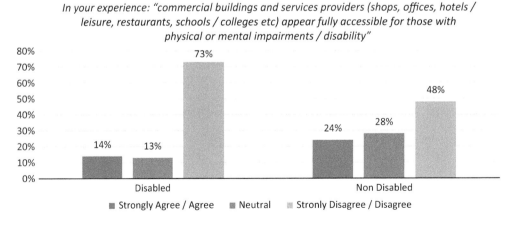

Figure 7.11 Accessibility of commercial buildings and service providers for the disabled v non-disabled.

those investing, advising on and operating commercial properties, goods and services to facilitate access. Despite this, the very nature of commercial real estate or enterprise is such that organisations will seek to establish the return on their investment.

Prior to discussing commercial attitudes to disability in the built environment, it is necessary to summarise the findings of the public attitudes research:

- **90%** of the respondents were aware of the legal requirement for service providers to undertake reasonable adjustment.
- **73%** of the respondents feel building owners should be responsible for funding reasonable adjustments to commercial properties.
- **73%** of the respondents who identify as having a physical or mental impairment **did not agree** that commercial buildings and service providers are fully accessible for those with physical or mental impairments.
- **73%** of the respondents aged 65 and over **did not agree** that commercial buildings and service providers are fully accessible for those with physical or mental impairments.
- **70%** of all respondents, both disabled and non-disabled visited commercial buildings 4–7 days in a week.
- **62%** of all respondents felt that funding is currently not sufficient in providing an inclusive environment.
- **48%** of the respondents who do not identify as having a physical or mental impairment disagreed that currently commercial buildings and services are fully accessible.
- **5%** of the respondents felt that end users or customers should be responsible for funding the costs or providing accessibility.
- **4%** of the respondents aged 65+ and recognising as having a physical or mental impairment visited commercial properties every day.
- **3%** of the respondents aged 65 and over visit commercial properties every day.

Commercial Attitudes Survey

The commercial approach and attitudes to disability and the built environment emanates from the opinions of those investing, advising on or operating commercial property, goods and services. In essence these primarily comprised:

- 70% property consultants;
- 14% investors;
- 11% service provider; and
- 6% real estate agents.

With approaches to disability initially shifting with the advent of the Disability Discrimination Act 1995, just over 40% of respondents to the commercial attitudes survey had over 20 years of experience. Accordingly, they will have seen the birth of the 95 Act and the revision in 2005, they will also be aware of its replacement by the Equality Act of 2010. Such an understanding is beneficial in contextualising the importance of early disability discrimination legislation.

The first significant point of note is that only 6% of the respondents who are all working age recognised as having a physical or mental impairment. This is significantly lower than the data provided by the government statistics, which equates to 22% over a whole-age population or 19% for the specific working-age group.

All commercial sectors were represented in the research data with 37% working in all property sectors. Concerning geographic representation, again, all regions of the UK are accounted for with approximately one quarter of the respondents working for firms with representation across the whole of the country.

Experiential Understanding of Accessibility

Somewhat reassuringly, the research data identified that 97% of those working or operating in commercial property are aware of the Equality Act 2010 and the requirement to provide access to goods and services.

Concerning an understanding of the term *Reasonable Adjustment* in the provision of access to goods and services for those with physical or mental impairments:

- 83% of respondents understand this term.
- 11% of respondents do not understand this term.
- 6% of respondents are unsure when it comes to understanding this term.

When asked to apply their experience in providing fully accessible buildings, relevant to their own field of operation, their responses are shown in Fig 7.12 and Fig 7.13.

Looking more closely at the current legislation (the Equality Act 2010), it is evident that only 40% of those owning, advising on, or providing goods and services in the commercial property sector agree that this is sufficient. Those unclear on whether the current legislation is sufficient at 29% is only marginally less than the 31% who feel it is

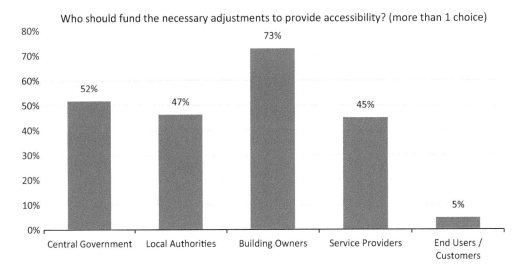

Figure 7.12 Funding for accessibility.

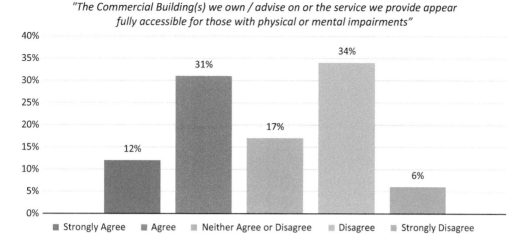

Figure 7.13 Accessibility of buildings/goods and services.

not sufficient. The apparent uncertainty and also mixed opinion on the suitability of the current legislation should act as a 'red flag'. With so few respondents recognised as having a physical or mental impairment, it means that their opinions are likely to be drawn from their education, training or experiencing disability from a surveying or implementation perspective.

Commercialisation of Accessibility

It has been established that the majority of those working within the commercial property sector are aware of the current legislation and the complexity of the application in the terms of 'reasonable adjustment'. However, there is less confidence that the current legislation is sufficient, and more importantly, whether their own commercial properties or business comply with this. Respondents were further asked to assess the statement that *There's commercial value in proving fully accessible buildings, goods and services,* and the overwhelming positive results are shown in Fig 7.14.

Concerning the possible sources of funding the implementation of accessibility, there is a resonance with the *Public Attitudes* in that nearly three quarters of those investing in, advising on, and operating commercial property believe building owners should be responsible. When pushed on the compliance with accessibility legislation of the commercial properties they are involved with, there is a slightly stronger emphasis on compliance to above the minimum legal standard as opposed to providing a standard to the legal minimum which is encouraging but ultimately there is an understanding that the levels of compliance vary (Fig 7.15).

Some key observations and themes emerging from the Commercial Attitudes survey are:

- the overall lack of property professionals and service providers with disability in the commercial property sector;

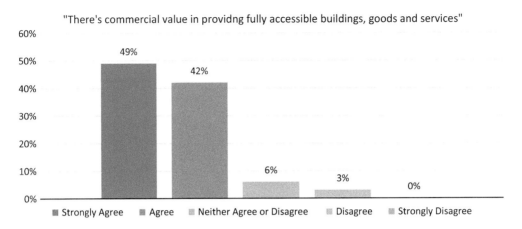

Figure 7.14 Commercial value in providing accessibility.

- the lack of experiential learning or understanding of disability of those driving design, implementing compliance or critically evaluating commercial property; and
- there appears a need for clarity in legislation and guidance notes on accessibility to deliver empathetic solutions by those who are not themselves disabled.

It is evident from the 'commercial attitudes' survey responses that there is a reasonable coverage of the UK with representation from UK wide regions accounting for about a quarter (23%) of the respondents. However, there is a strong bias towards investors, owners, consultants, agents and service providers operating in London and the Southeast of the UK. Regional representation is perhaps more relative to areas of economic prosperity. This may skew the findings that there is slight inclination for implementing over legal minimum standards. Despite this, there is an overwhelming recognition that

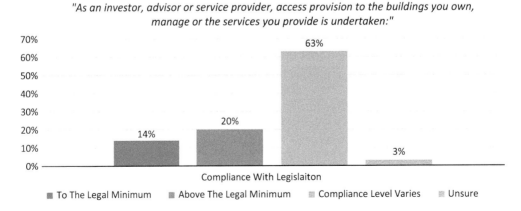

Figure 7.15 Current levels of access provision.

full accessibility has commercial value, with nine out of ten respondents expressing this view. However, when asked to assess the current levels of accessibility to the built environment and also the suitability of the current legislation, it is evident that this is much more inconclusive. Accordingly, there appears a disconnect between the potential and actual situations.

The key findings of the data analysis into Commercial Attitudes to Disability in the Built Environment can be summarised by the following key numbers:

- **97%** of the respondents **were** aware of the requirements under the Equality Act 2010 to provide access to goods and services.
- **91%** of the respondents **agreed** that there is commercial value in providing accessibility.
- **83%** of the respondents **understand** the term *Reasonable Adjustment* in the provision of accessibility.
- **74%** or the respondents believe **building owners should be responsible** for funding the necessary reasonable adjustments to facilitate accessibility.
- **63%** of the respondents **confirm that the level of access provision varies** across the buildings they own, manage, advise upon or the services they provide.
- **54%** of the respondents believe that **service providers should be responsible** for funding the necessary reasonable adjustments to facilitate accessibility.
- **49%** of the respondents believe that **local authorities should be responsible** for funding the necessary reasonable adjustments to facilitate accessibility.
- **40%** of the respondents **agree that the current legislation concerning accessibility is sufficient.**
- **40%** of the respondents **disagree** that the buildings they own and advice upon or services they provide are **fully accessible.**
- **31%** of the respondents **disagree that the current legislation concerning accessibility is sufficient.**
- **29%** of the respondents are **unsure that the current legislation concerning accessibility is sufficient.**
- **26%** of the respondents feel that end users / customers should be responsible for the funding of accessibility.
- **20%** of the respondents **confirm minimum level of compliance for access provision** to the buildings they own, manage and advise upon or the services they provide.
- **17%** of the respondents are unsure or do not understand the term *Reasonable Adjustment* in the provision of accessibility.
- **14%** of the respondents **confirm above minimum level of compliance for access provision** to the buildings they own, manage and advise upon or the services they provide.
- **Only 6%** of the respondents involved in investing, owning and advising on commercial property or providing goods and services identify as having a disability.

Contextualised Public and Commercial Attitudes

Concerning the current legislation, the 'headline' data from this research detailing the legal prescription is that both commercial (**97%**) and public attitudes survey (**90%**) respondents are aware of The Equality Act 2010. Of note is the awareness of the requirement of service providers to undertake reasonable adjustment. This perhaps dispels the notion that repealing the Disability Discrimination Act 2005 and replacing it with the Equality Act 2010 has meant a loss of focus on disability, as detailed in the article by David Blunkett. It should however be noted that in the in the terms of legal challenge brought under the Equality Act and in particular access to goods or services, there has been little in the way of high-profile cases. The exception to this is *Paulley v First Group plc,* but this did not specifically concern access to the built environment, so it is not clear how effective the Equality Act 2010 is concerning enforcement procedure for commercial properties or service providers.

Property professionals and service providers have an obligation to have knowledge of the legislation. Their understanding is likely to be reinforced if they have undertaken a higher education qualification within the built environment disciplines where accessibility is embedded in the courses accredited by the Royal Institution of Chartered Surveyors (RICS) or Chartered Institute of Building (CIOB). This is the case with those who identify as being consultants or agents in the research data and is perhaps is the reason why there is a 97% awareness of the Equality Act 2010 and the need for reasonable adjustment.

Awareness of the legislation and the obligations placed on service providers is also evident in the findings of the 'public attitudes' survey. The level of awareness from the end users is encouraging in the greater context of inclusion in society and may be a result of the increased publicity centred around disability associated with events such as the Paralympics. However, technically an understanding of the application of 'reasonable adjustment' is likely to be less well understood as the general public are unlikely to have undertaken specific research, training or education on this subject. This is reflected by the data that nearly one quarter of respondents identified in the 'public attitudes' research data neither agree or disagree that commercial properties and service providers are fully accessible to those with a physical or mental impairment.

Furthermore, one quarter of respondents to the 'commercial attitudes' survey also expressed a neutral opinion concerning the accessibility of commercial buildings. This suggests that even with specific research, knowledge, training or education on the subject, it is still difficult for some to assess whether commercial properties, goods and services are fully accessible.

Visiting Commercial Properties or Accessing Goods and Services

Having analysed the research data concerning visits undertaken to commercial properties, younger (working age) respondents and those without disability do this with more frequency than those with a physical or mental impairment. Those aged above 65 and those also in this age category identifying as having a physical or mental impairment are less likely to make daily visits. However, approximately three quarters of all respondents (**73%**) made visits to commercial properties 4–7 times a week. Therefore, it can be concluded that a significant majority of the 'public attitudes' survey respondents experience, on a regular basis, accessing commercial buildings, goods and services. This is an important consideration as respondents were asked their opinion on the current accessibility regimes of commercial properties or services for those with a physical or mental impairment.

The Perception Gap

Based on their experience of accessing commercial properties or goods and services, there is a significant difference of opinion between those with disability and those without. Nearly three quarters (**73%**) of those with a physical or mental impairment do not agree that commercial buildings or goods and services are fully accessible. This differs from nearly half (**48%**) of those without disability who also disagree with the same statement. This represents a **25%** gap between the two opinions and is comparable, to an extent, with the *Scope Disability perception gap – report* published in 2018. This report detailed a 'gap' of 10% in the perception of discrimination felt by those with a disability compared to the perceptions of disability discrimination by non-disabled people. In the context of this research and a 'gap' of 25% in perception on the accessibility of commercial properties, this is likely to be heightened or magnified by a tangible experience. This can only truly occur to those with a disability or those caring for a disabled person when accessing goods or services. The implications of the perception gap detailed in the *Scope* report includes a suggestion that this contributes to an increase in discrimination. Consequently, if that theory is applied to this research, together with evidence of a wider gap, it may result in reduced awareness and support for improved access to commercial buildings.

There is a significant deviation from the initial findings that 90% of the public have an awareness of the Equality Act 2010 and the term 'reasonable adjustment' compared with their experience of accessibility. When examined in more detail to establish whether current commercial properties are fully accessible, this produces a perception gap. It perhaps is best embodied by the sentiment following the London 2012 Paralympics. The UK Government claimed an improvement in the public perception of disability (Gov.UK, 2014), which differs significantly to the opinions of both Paralympian Sophie Christiansen (2013) and that published by Scope (2018). It is obvious that despite an overwhelming majority being aware of the measures to prevent disability discrimination, there is a lack of understanding from non-disabled respondents on actual accessibility which appears to mirror the contradiction in published work pre and post 2012 Paralympics.

The Public and Commercial Attitudes survey also interviewed two key individuals representing a disability charity and also involved in strategic advice on accessibility. The responses of the interviewees concerning the perception gap effectively questions whether the minimum standards are really accessible. There is a feeling that there is a fundamental difference between what is accepted under Approved Document Part M of the Building Regulations and what is actually accessible in practice. It is also evident that compliance with building regulations does not indicate compliance with the Equality Act 2010, as well as an ignorance concerning the legislation and regulations. There is a sense that there is an unwillingness to accept that discrimination is very common and institutionalised. These are strong opinions emanating from both participants' extensive experience of working closely with or representing those with a disability, suggesting more a deep-set misunderstanding of disability and the legal frameworks.

Compliance – Commercial Attitudes and Public Perception

There is an overwhelming **97%** awareness from investors, owners, consultants, agents and service providers of the Equality Act 2010. As previously indicated, those operating within the commercial sector and in particular those delivering advice, e.g., agents and/ or consultants, are expected to have more than just awareness of the legislation. Although they are not necessarily expected to be specifically trained *access consultants.* They should be able to identify potential areas of non-compliance or building issues that might affect accessibility, accordingly, 83% of respondents understand the term 'reasonable adjustment'. The findings of the research data established varied opinions regarding the accessibility of buildings, goods and services. There is a marginal majority of respondents (**43%**) who feel the buildings they own, advise on, or the goods and services they deliver are fully accessible to those with physical or mental impairments. In contrast, **40%** of respondents from the 'commercial attitudes' survey disagrees and **17%** have a neutral opinion.

The overall response rate from the 'commercial attitudes' survey was lacking in the opinions of agents and actual service providers, so it is difficult to draw significant conclusions about their opinions. However, amongst consultants, opinion is divided on whether the properties they advise upon are fully accessible, with 36% agreeing and 36% disagreeing with the statement. Consultants are likely to be the most 'qualified' in the delivery of advice on accessibility. Their education, professional obligation and experience of commercial property means they are often instructed to deliver evidence-based opinion on accessibility in the course of their work.

There is a significant contrast between the public attitudes and commercial attitudes surveys on whether commercial properties or goods and services are fully accessible. The perception gap identified in previously widens further when the responses of the 'commercial attitudes' survey are included as in Fig 7.16.

Considering the lack of commercial respondents who identify as having a physical or mental impairment (**6%**), it is likely that knowledge of accessibility is derived from the legal framework taught as part of education, training or guidance documents as opposed to an *experiential understanding*. The actions associated with performing surveys of commercial buildings and in particular access audits, compliance with legal standards

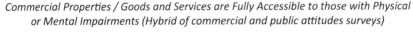

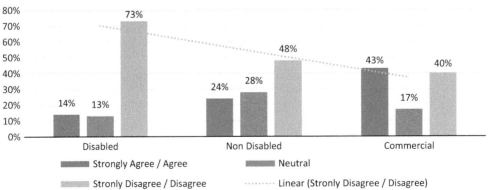

Figure 7.16 The 'perception gap' concerning the accessibility of commercial properties, goods and services.

should be binary, it either complies or it does not, as discussed in Chapter 6. However, 'reasonable adjustment' is a term that can be used to 'soften' or compromise access obligations where it is physically impossible to do this within the existing built environment. Those working in commercial property are more likely to adopt a pragmatic application of the law regarding reasonable adjustment. This may be one significant reason why 40% of respondents disagree with the statement that the commercial buildings they advise on are fully accessible.

It has been acknowledged that there exists a 'perception' gap between disabled and non-disabled respondents in the 'public attitudes' survey around whether they feel buildings, goods and services are fully accessible. This can be reasoned to a lack of experiential understanding where those without disability feel differently than those with a physical or mental impairment. This does not explain why those operating within the commercial property sector or the delivery of goods and services have such a significant difference of opinion. Disability is an emotive subject, while legal prescription concerning the access of goods and services is objective, it does not replace the experience or struggles of actually attempting to gain access to buildings, goods and services. There appears a difference in the understanding of disability access provision with knowledge-based opinion of the 'commercial attitudes' respondents contrasting significantly from practical opinions of those with disability. It is therefore necessary to establish whether the legal prescription and guidance is sufficient to establish ways to bridge this gap in understanding.

Concerning the suitability of the current legislation, it has been noted that the response from the 'commercial attitudes' survey has identified that 40% of respondents believe that this is suitable, although there are 21% who disagree. Considering that the majority of the commercial attitudes' respondents do not identify as having a physical or mental impairment, it is not clear why only 40% believe the current legislation is sufficient. The respondents are likely to be more than aware and, in most cases, formally taught or trained in the concept of accessibility. Their willingness to doubt the sufficiency of current legislation may come from a genuine belief in a requirement for improved accessibility. Alternatively, it may be based on either their experience of delivering goods and services or advising property investors, owners and occupiers. It is not clear if the answers to this question are objective and governed by the legal prescriptions or subjective and 'feelings' based. Over half (**57%**) of the respondents who disagree that the current legislation is sufficient also disagree that the buildings they advise upon or the goods and services they deliver are fully accessible. This establishes a limited correlation but more importantly begins to question the appropriateness of the existing legal provision or guidelines on accessibility.

It can therefore be suggested that there is a disconnect between the theory based legal prescription; which includes the legal minimum requirements and the practical application of those experiencing disability. There seems an obvious requirement to 'close' this gap. The options for doing this include:

- analysing the effectiveness of the existing minimum legal requirements;
- upgrading or strengthening the minimum legal requirements;
- consulting those with physical or mental impairments on areas of specific access concerns and their experiential difficulties with access; and

- developing a more comprehensive accessibility grading or rating system and communicating this to the wider general public this should include those with disabilities to provide simple and effective communication about where and how goods and services can be fully accessed.

Levels of Compliance

When considering the levels of compliance for accessibility, the default and minimum requirement is by adopting the content of Approved Document Part M of the Building Regulations; however, it is important to note that for existing buildings this is not always enforceable but provides a benchmark or reference set of requirements. Under the Equality Act 2010, it is necessary to undertake 'reasonable adjustment' in existing buildings to avoid discrimination and this should be done wherever possible to meet the legal minimum requirement. Above the minimum requirement compliance commonly equates to access provision, which is drawn from all or part of British Standard 8300 (BS8300, 2018) Parts 1 and 2. Responses to the 'commercial attitudes' survey confirmed that for **63%** of respondents, the level of compliance varies between the legal minimum and above the legal minimum. This reflects the varied nature of the existing built environment, as well as the associated complexity in implementing access provision. A small number of respondents (**14%**) confirmed that levels of compliance were above the legal minimum, which illustrates that one in 7 respondents are making this choice, but the constraints of the existing built environment often dictate compromise. Therefore, the combined survey data is that **77%** of respondents undertake a mixture of legal minimal and above minimum compliance and that improved access is achieved because they want to and not because they have to. The positive nature of creating above minimum requirement reasonable adjustment to meet the requirements of the Equality Act 2010 contradicts the negative commentary of Hand et al (2012) as well as fact sheet produced by Unison in 2015 commenting on 20 years of disability discrimination legislation.[14] It was their assertion that with the Equality Act 2010, this 'new' piece of legislation would revoke much of the change initiated by the Disability Discrimination Act 1995. However, the findings of this research dispel this notion, but it still does not explain the perception gap between disabled users and commercial providers on the basic accessibility of commercial properties.

Contradicting the survey findings, it is evident in the responses of the interviewees that investors, property owners and service providers going above the minimum standard is the exception. There appears a willingness to revert to the basic minimum standards in Part M as an acceptable status when this is not always accessible. There is a uniform criticism that the current UK legislation does not equate to a set of building codes, such as that used in the USA; added to this is that the existing building regulations are not being enforced. Furthermore, the interviewees suggest compliance with the minimum standards does not always mean that buildings are fully accessible and this is an opinion widely shared by the inclusive design community.

Accessibility – Use of Commercial Properties and Access to Goods and Services

The headline data from the 'public attitudes' survey is that almost three quarters of respondents make visits to commercial properties 4–7 days a week. Accordingly, it appears that service providers have a vested interest in being accessible as well as being open. There is a broad trend in the number of visits decreasing with age, which is illustrated by the data that only **3%** of those age 65 and over are visiting commercial properties every day. There is a similar trend evident with those who identify as having a disability in that they also visit commercial properties with less frequency. However, it should be acknowledged that visit frequency may be linked to a lack of an accessible transport systems and this is something which was not part of the research.

There is a link in the survey data between age and disability with the over 65s recognised as having increased physical or mental impairments, the prevalence of age-related disability is similar to data published by the UK government in 2018. The following characteristics have been established concerning the survey population of those over 65:

- increased incidence of physical or mental impairments compared to those of working age.
- **78%** of over 65s who identify as having a disability, indicate this to include mobility as an issue.
- **42%** of over 65s who identify as having a disability indicate this to include visual impairment as a disability.
- Over 65s visit commercial properties or service providers with lesser frequency than those of working age.

Perception on Accessibility

A general trend has been established that a majority of those with physical impairments including mobility, stamina, breathing, dexterity, vision and hearing have views that commercial properties, goods and services are not fully accessible. This is magnified significantly for the over 65s where impairment is more prevalent.

It has been established that 73% of respondents over the age of 65 and 73% of respondents who identify as having a disability, disagree when drawing from their own experience that commercial buildings, goods and services are fully accessible. Deeper analysis of the data for the over 65s per category of disability has identified:

- **73%** of those who identify **mobility** as their physical impairment disagree that commercial buildings, goods and services are fully accessible; and
- **75%** of those who identify **vision** as their physical impairment disagree that commercial buildings, good and services are fully accessible.

This establishes that the two largest demographics of those with disability who do not feel commercial properties, goods or services are fully accessible are the **over 65s with mobility issues and visual impairment**.

Commercial Value and Funding of Inclusivity

The findings of the 'commercial attitudes' survey is that **91%** of respondents strongly agree or agree that there is **commercial value in** providing fully **accessible buildings**, goods and services. This correlates with the findings if the Accenture report published in

2018, which also mentions engagement, enabling and empowering those with a disability from a commercial perspective. The concept of empowering those with a disability is also discussed in an article authored by David Blunkett in 2015, and parallel to the commercial value of providing accessibility is the social benefit of supporting disabled people to contribute to their own communities. The missing link between the recognition of disability within the built environment and achieving the commercial and social value of this is likely to be caused by the perceived costs associated with making reasonable adjustments.

Implementation of Accessibility

The Equality Act 2010 obliges service providers to undertake reasonable adjustment to prevent discrimination against those with physical or mental impairment seeking access to goods and services. The majority of reasonable adjustments will involve removal or alteration of building features or the provision of alternative access options. In most cases, the service provider within commercial properties is either an owner occupier or tenant. In the event that the service provider is a tenant, then alteration of the property and in particular removing or altering building features will likely be governed by the nature of their lease agreement. Despite this, the response of both the commercial and public attitudes surveys identified that the 73% and 74% or the respective respondents felt that building owners should be responsible for funding the necessary adjustments.

Commercially the costs of implementing access requirements may be absorbed by the building owner or service provider, depending on the value of the works, size of the organisation and available funds. It is possible that the costs may be passed on to the tenants or service providers and eventually customers; however, the majority (91%) opinion is that there is commercial value in providing accessibility. There appears to be an established incentive for building owners and service providers to invest in accessibility with 22% of the population recognising as having a physical or mental impairment. Despite the over 65s and those with disability undertaking less frequent visits to commercial properties, more emphasis should be placed on the quality and not quantity of their means of accessing goods and services.

When asked to consider the principal barriers and drivers for inclusivity, interviewees suggested a need for a change in attitude away from minimum standards culture. Given the government's role in delivering, in conjunction with private sector vast infrastructure projects, the government can lead by example was the opinion of one of the participants. Furthermore, the participant went on to recommend the notion to make inclusive design a requirement of all aspects of publicly funded works and hence encompass the sense that inclusivity is a benefit for all as opposed to a minority. More could also be done to embrace access consultants in the process as well as working with end users was the opinion of the other participant. There is a notion that costs associated with inclusivity present a barrier, but it is more cost effective or cost neutral to implement this at the design stage.

There is an urgent need to strengthen legislation (both the Equality Act 2010 and the Planning and Building Act) as well as ensuring much better enforcement of the existing standards and a need for trained *Approved Inspectors*. Proper enforcement and the existing legislation as well as the provision of state-funded legal aid to allow disabled people to challenge access provision is also something that could yield more positive results. Such opinions are evident from those at the 'sharp end' who have been either representing disabled people or working on the application and implementation of inclusive design. There is an apparent social, attitudinal and commercial resistance to change.

Strengthening legal controls goes against opportunity and in favour of obligation, its choses 'stick' over 'carrot', which is confrontational in an economic environment where the public purse is stretched and the possibility of legal aid practically non-existent. The cash reserves in commercial property, perceived commercial value of inclusivity and the benefit over cost may be a better way to structure the 'argument'.

While the commercial value attributable to fully accessible commercial properties has been established in the survey results, the interviewees took a more holistic approach. The social value of inclusivity is not to be underestimated with socially inclusive companies being more productive along with the added commercial value seeing disabled people as:

> ... .citizens with spending power and also as excellent employees.

The following recommendations should be adopted to ensure reasonable access to commercial buildings, goods and services:

- grading and marketing or commercial properties, goods or services which are fully accessible to allow users to make a choice;
- establishing what can be gained or the commercial 'edge' achieved through the provision of fully accessible goods and services;
- analysing the cost benefit of short-term investment against commercial gain through the adoption of accessible environments; and
- encouraging the adoption of an above legal minimum compliance level to showcase and replace the feelings of 'obligation' often associated with legal compliance to one of commercial 'opportunity'.

Funding

The majority of responses to the commercial and public attitudes surveys indicate building owners and service providers as those responsible for funding the necessary adjustments. The minority responses were for the end users or customers to pay for this with 26% of commercial attitudes responses suggesting this compared to 5% of public attitudes responses. The notion of 26% of respondents suggesting 'end users' pay could be a misunderstanding in commercial terminology where the tenant is often seen as the end user.

The interviewees agree that it is wholly inappropriate for those with a physical or mental impairment to pay extra for goods and services, this in essence contradicts the very core of the Equality Act 2010 and is illegal. It is recognised that living with a disability incurs more cost, and in essence, disabled people are simply fed up with having to pay more for things that are a basic right. Tax relief or tax credits can be considered as a way to 'reward' commercial organisations for implementing adjustments to ensure accessibility; however, this feels inappropriate when it should be the 'norm' to undertake these works and not seen as something to 'reward'.

There is a feeling amongst disabled people, organisations representing disabled people and those working to remove physical barriers and create an inclusive built environment that discrimination is endemic and institutionalised. Citing the "Black Lives Matter" campaigners, there is a sense of anger and frustration that accessibility is only given 'lip service' by those in power. It goes unrecognised or is ignored by commercial concerns, so a substantial section of our community continues to be excluded. This state of affairs should not be tolerated any longer.

Reflecting on Public and Commercial Attitudes to Disability in the Built Environment

Reflecting on the findings of the research data and the emerging themes from the review of the existing body of knowledge the following findings have emerged from the research undertaken into public and commercial attitudes:

- There is an overwhelming awareness of the Equality Act 2010 by both public and commercial respondents in respect of the obligation to make goods and services available to those with physical or mental impairments.
- The commercial attitudes survey has identified that those advising on commercial property or providing goods and services neither agree nor disagree that the current legislation is sufficient.
- In line with the report published by Scope: *Disability Perception Gap – Report* there is also a significant difference in how those with a disability, those without a disability and property professionals / service providers view the current accessibility of commercial properties. Importantly, there is an obvious difference between the theory from legislation and guidance documents and practical application, gained from experiential understanding regarding current access provision.
- There is a suggestion that investors, owners, consultants and service providers are implementing some above minimum legal requirement compliance. This clearly does not appear to be experienced by those with a physical or mental impairment, as detailed by the perception gap identified in this research. It has also been questioned whether the minimum standards are sufficient as there is evidence of buildings that are deemed to be compliant, but in reality, are not accessible.
- There is an overwhelming feeling that there is commercial value in providing accessibility to the built environment with an emphasis on building owners as well as service providers to fund the necessary adjustments. Although there should be a push to change attitudes or evolve processes away from the 'check box' approach of minimum compliance. Obvious commercial value can be seen in the terms of income or profit; however, true value is both financial and social with much longer lasting or generational benefit.
- Living with a physical or mental impairment does incur more costs. This may be a contributing factor why those with a disability are less willing to pay or contribute when they also consider this to be a basic right.

Promoting Opportunity Over Obligation, 'Carrot over Stick'

There should be no reason why access provision should not be considered by all commercial property owners and service providers. Ultimately, there is commercial interest and value in being wholly accessible. The research has identified that primarily building owners as well as service providers should be responsible for providing the funding to facilitate access. It should not be the responsibility of the taxpayer to fund this as, apart from the feel-good factor associated with an inclusive environment, it is the service providers and customers who ultimately benefit from enhanced access provision. Suggesting that end users are obliged to pay 'extra' for the same goods and services contradicts the principles of the Equality Act 2010.

Commercial organisations invest heavily in corporate social responsibility, but this is probably not something considered by small businesses or individual service providers. However, for relatively modest investments in accessibility, it is possible for all commercial service providers to reap the rewards commercially. This could be recognised with the certification or accreditation of those organisations choosing to contribute to an accessible built environment by making this a strategy as part of their corporate social responsibility.

Having established that there is commercial value in providing an inclusive environment and commercial organisations should seek to choose opportunity over obligation. However, there is a need to undertake a reality check in the sense that asides from implementing inclusive design to new build properties, the majority of access solutions will be a retrospective action. Accordingly, this is further complicated with the implementation to historic buildings. Chapter 8 will detail the challenges and legal prescriptions associated with this and Chapter 9 will illustrate the application of legal minimum standards as well as best practice in creating a fully inclusive environment.

Notes

1 Paralympics History – Evolution of the Paralympic Movement
2 Public-and-Commercial-Attitudes-to-Disability-in-the-BE.pdf (ucem.ac.uk)
3 Building for equality | Construction Industry Council (cic.org.uk)
4 Building for Equality: Disability and the Built Environment (parliament.uk)
5 Tagg. 2018. Technical Due Diligence and Building Surveying for Commercial Property. Routledge.
6 BPF – About Real Estate
7 Death of the high street weighs on landlords round the world | Financial Times
8 Is this the high street's last stand? | Financial Times (ft.com)
9 Death of the High Street is overstated, claims Which?, as analysis shows rebirth of independent shops (telegraph.co.uk)
10 The Accessibility Advantage: Why businesses should care about inclusive design (pfp-idefellowship.org)
11 what-workers-want-uk—2019.pdf (savills.co.uk)
12 Disabled people in employment (parliament.uk)
13 Public-and-Commercial-Attitudes-to-Disability-in-the-BE.pdf (ucem.ac.uk)
14 Cuts threaten disabled people in work | Article, News | News | UNISON National

References

Accenture. (2018). *The accessibility advantage: Why businesses should care about inclusive design*. [online] Available at: https://pfp-idefellowship.org/wp-content/uploads/2018/12/Accessibility-Advantage_PoV_FINAL.pdf. [Accessed 14 January 2024].
Bsigroup.com. (2018). BS 8300-2:2018. Design of an accessible and inclusive built environment. Buildings. Code of practice. [online] Available at: https://shop.bsigroup.com/ProductDetail/?pid=000000000030335835 [Accessed 14 January 2024].

Christiansen, S. (2013). A year after the Paralympics attitudes to disability need to improve. [online] Available at: https://www.theguardian.com/sport/blog/2013/aug/24/paralympics-sophie-christiansen-equestrian [Accessed 14 January 2024].

Department of Work and Pension. (2018). Family Resource Survey 2016/7 [online] Available at: https://assets.publishing.service.gov.uk/government/uploads/system/uploads/attachment_data/file/692771/family-resourcessurvey-2016-17.pdf [Accessed 25 November 2019].

Dixon, Smith and Touchet. (2018). *Disability perception gap – Policy report*. Scope. [online] Available at: https://www.scope.org.uk/campaigns/disability-perception-gap/ [Accessed 14 January 2024].

Evans. (2019). *Death of the high street weighs on landlords round the World*. [online] Available at: https://www.ft.com/content/4e5e1022-8df3-11e9-a1c1-51bf8f989972. [Accessed 14 January 2024].

Financial Times. (n.d). *Is this the high street's last stand?* [online] Available at: https://www.ft.com/content/946e60d0-b75b-410c-9d0f-f52f0ab35c89. [Accessed 14 January 2024].

Gov.UK. (n.d). The Equality Act 2010. [online] Available at: https://www.legislation.gov.uk/ukpga/2010/15/contents [Accessed 14 January 2024].

Gov.uk. (2014). 'Transformation' in British attitudes towards disabled people since Paralympics 2012 – GOV.UK. [online] Available at: https://www.gov.uk/government/news/transformation-in-british-attitudes-towardsdisabled-people-since-paralympics-2012 [Accessed 14 January 2024].

Hand, J., Davis, B. and Feast, P. 2012. Unification, simplification, amplification? An analysis of aspects of the British Equality Act 2010. Routledge [online] Available at: https://www.tandfonline.com/doi/abs/10.1080/03050718.2012.695001 [Accessed 14 January 2024].

House of Commons Library. (2023). *Disabled people in employment*. [online] Available at: https://researchbriefings.files.parliament.uk/documents/CBP-7540/CBP-7540.pdf [Accessed 14 January 2024].

House of Commons Women and Equalities Committee. (2017). *Building for equality: Disability and the built environment*. [online] Available at: https://publications.parliament.uk/pa/cm201617/cmselect/cmwomeq/631/631.pdf [Accessed 14 January 2024].

International Paralympic Committee. (n.d). *Paralympics history*. [online] Available at: https://www.paralympic.org/ipc/history#:~:text=On%2029%20July%201948%2C%20the,who%20took%20part%20in%20archery. [Accessed 14 January 2024].

Johnson. (2019). *Death of the high street is overstated, claims Which?, as analysis shows rebirth of independent shops*. [online] Available at: https://www.telegraph.co.uk/news/2019/10/19/death-high-street-overstated-claims-analysis-shows-rebirth-independent/. [Accessed 14 January 2024].

Savills. (2019). *What workers want: UK*. [online] Available at: https://pdf.euro.savills.co.uk/uk/commercial—other/what-workers-want-uk—2019.pdf [Accessed 14 January 2024].

Tagg. (2018). *Technical Due Diligence and Building Surveying for Commercial Property*. Routledge

Tagg. (2020). *Public and commercial attitudes to disability in the built environment*. [online] Available at: https://www.ucem.ac.uk/wp-content/uploads/2020/10/Public-and-Commercial-Attitudes-to-Disability-in-the-BE.pdf [Accessed 14 January 2024].

The British Property Federation. (n.d). *About real estate*. [online] Available at: https://bpf.org.uk/about-real-estate/ [Accessed 14 January 2024].

The Construction Industry Council. (2017). *Building for equality*. [online] Available at: https://www.cic.org.uk/news/building-for-equality?s=2017-04-25-building-for-equality [Accessed 14 January 2024].

Unison. (2015). *Cuts threaten disabled people in work*. [online] Available at: https://www.unison.org.uk/news/article/2015/10/cuts-threaten-disabled-people-in-work/ [Accessed 14 January 2024].

Chapter 8

Accessing the Historic Built Environment

Introduction

For some, there is a romance or charm associated with old buildings, and these can be great places to live, work or socialise. However, the obvious negative connotation associated with the Historic Built Environment (HBE) is that this was constructed prior to any significant legislation. There is no doubt that these in many instances have effectively stood the 'test of time' and what might be seen by some as 'character', 'charm' or 'quirk' is experienced by others with disability as inconvenient or simply inaccessible. Examples of this may be too-narrow doorways, sloped uneven floors or the absence of lifts that prevent vertical circulation. These are just a few examples of the wide variety of accessible requirements that need to be considered when providing an inclusive environment. The notion of 'reasonable adjustment' has already been discussed and observed to deliver compromises on accessibility. This is made significantly more difficult with heritage buildings as they are often afforded legal protection meaning that cannot be readily changed without specific permission.

While it is accepted that there is public awareness of buildings with historical or cultural value such as castles, churches, cathedrals, palaces or museums, there are thousands of 'regular' commercial buildings that are also protected, such as shops, pubs and hotels. This places restriction and limitation on what can be done in the terms of delivering accessibility and appears to engender a greater emphasis on compromise. Added to this are a number of organisations as well as individuals with an active interest preserving historic buildings, which can evoke some strong opinions. To contextualise the issues associated with implementing access to historic buildings, it is necessary to define heritage and the historic built environment.

Heritage and the Historic Built Environment

To begin the process of understanding the access challenges associated with the Historic Built Environment, it may be appropriate to ask the question why this is protected. If it is all about accessibility as well as clean, modern and energy-efficient buildings, then surely there is no place for heritage property in modern society? However, this could not be further from the truth with examples of landmark historic buildings that represent global or national identity. To most people, almost every country has an iconic building or monument that symbolises a moment in time or represents a historical event. For some, these do not always represent good events but could equate to social struggle,

DOI: 10.1201/9781003239901-8

persecution or conflict. There is a school of thought that we cannot or should not try to erase history but instead try to learn from it; such a simple statement does not give credence to those directly or indirectly related to the historical events or the nuances of this which may have been lost in the passage of time. In many ways, buildings are often the only tangible link between our past and the present.

While the presence of global landmark buildings may shape or define national identity, this is mirrored with local heritage buildings which represent vernacular existence; in the UK, this spreads from South to North and East to West. It is represented by places of worship, industry, residence and community. These buildings are composed of the external envelope, fabric and structure forming the carcass alongside the fit out, fixtures and finishes that places flesh on the bones of history. We can all look around us in every location in the UK and identify local buildings that form part of the social narrative or story where we live or come from. Quite frankly, once historical buildings are significantly altered or even demolished, this piece of history is erased forever and cannot be replaced. There is no cloning process for unique heritage buildings, and the DNA of these not only comprises those who designed and built them but also those who occupied and used them. The owner, occupier or service provider for every heritage building is temporary, albeit their residency could last decades commercially or even centuries in the context of the royal palaces. They are custodians of the building fabric and history, and the changes or alterations they make will be passed onto the next generation of owner, occupier or service provider. Quite simply, once it's gone, it's gone.

Another key consideration or benefit of preserving the historic built environment is the economic value of this as defined by Historic England as the Gross Value Added (GVA) of the sector. In essence, GVA is a measure of the economic value that owning, leasing, occupying, repairing and visiting of heritage assets. In real terms, Historic England estimate this accounts for over half a million jobs with a combined economic value of £36.6bn. The report published in 2020 contains significant forensic analysis to evidence base the findings. Some of the granular detail notes that investing in the historic built environment increases footfall and reduces vacancy rates. One of the key statistics emanating from the report is that more than 25% of listed buildings are occupied and operated for commercial use, in real terms, this estimated to be approximately 142,000 properties.

Therefore, it is easy to understand the argument for building preservation from a historical or cultural perspective, to examine this further, it is necessary look at the long-established discussion on building preservation versus building conservation. Historic England are perhaps one of the principal organisations to consult when looking at the notion of conserving the historic built environment, and they have their own distinct definition of conservation.[1] In essence, their definition of conservation identifies this as the necessary process to manage change to a heritage asset in the context of sustaining as well as enhancing its significance. Therefore, undertaking a retro upgrade of such an asset can be managed to sustain the historical value, and surely by making this accessible, it enhances the significance of this on the basis that is had been made accessible? This definition appears to be open to interpretation, similar to the very notion of 'reasonable adjustment'. In contrast to the concept of conservation established by Historic England is that of the Society for the Protection of Ancient Buildings (SPAB). It was formed in 1877 by William Morris to combat the overenthusiastic approach of Victorian architects in their attempts to undertake restoration which was deemed as harmful.[2] SPAB appear very much more focussed on 'protection' with the building fabric being at the heart of this. While SPAB also refer to conservation, it can be interpretated by their approach that they sit more towards the preservation side of the conservation curve. This is exemplified by their stance against 'restorationist arguments',[3] which would appear to challenge any permanent approach to providing accessibility as this would invariably result in alteration to the fabric of a building.

When undertaking audits and advising on accessibility the principle governing retro fit out is that of 'reasonable adjustment', and it has been discussed the notion that reasonable historically favours the service provider or employer. Compounded with this is a sense that any retro fit works must not be of detriment to the heritage value of a building, which implements another barrier to full inclusivity. However, it is also important to refer back to the wording of Historic England and the further notion that change can be made provided it both sustains and enhances its heritage value. It is also necessary to put this into hierarchical context relative the historical or cultural importance of a building.

Presently and according to Historic England, there are approximately 500,000 listed buildings in the England,[4] although some of these represent groups of buildings which make up a single entry on the list. The variety of the relevant architecture is very diverse and a listing may relate to something as simple as a monument or small single dwelling through to vast royal palace or even sports stadia. There is above all else a need to appreciate the existence of the classification of listing in order of importance:

- Grade 1 – Exceptional Interest (2.5% of listed buildings);
- Grade 2* – Particular importance (5.8% of listed buildings); and
- Grade 2 – Special interest (91.7%).

The classification of the listing certainly impacts upon any conditions imposed upon the building owner in regards to maintenance, repair and alteration. A building may not be listed but may form part of a collective of buildings or streetscape that can also be given a degree of protection by designating this a Conservation Area. To prevent inappropriate works or even demolition occurring to a listed building or even works within a conservation area, there is legislation in place to govern this.

Legislation

Established in 1990 the Planning (Listed Buildings and Conservation Areas) Act 1990 is the principal legislation use to protect or conserve the historic built environment. Buildings are selected for listing based upon principles published by the Department of Culture, Media and Sport.[5] This guidance is publicly available and when considering listed status, the secretary of state responsible for this government department will consider some of the following key criteria:

- architectural interest;
- historical interest;
- group value;
- fixtures and features of a building & curtilage buildings; and
- the character or appearance of conservation areas.

Figure 8.1 A Grade 2* timber framed 14th or 15th century property, typical of that associated with 'listed' status with contemporary use.

The published guidance also identifies a number of general principles that state the age and rarity are key. As might be expected, buildings pre-dating 1700 are almost certainly included as are a selected number of buildings from the 18th to late 19th centuries.

In the context of providing access to commercial property, goods, services and employment, one obvious application of listed buildings originating from the mid to late 1800s are low-rise retail properties associated with shops, restaurants and pubs across the town centres of the UK (Fig 8.1).

While modern buildings or those of less than 30 years of age are not normally selected for listing, there exists a case to afford this listed status if they are examples of specific construction technology or are associated with key historical events (Fig 8.2).

While the two examples of listed buildings are separated by 2–3 centuries in their date of construction and appear dramatically different in all aspects of their design, they share two common characteristics. Primarily, they are both protected by the same legislation to prevent unauthorised alteration and irrespective of the age difference, and they were both designed without the presence of disability specific legislation. While both buildings have current different destination use, it should be noted that the service provider for each is also obliged to make reasonable adjustment to facilitate access to goods, services and employment. This exemplifies the positioning of blanket legislation which requires very individual, contextualised application to deliver appropriate design solutions.

While it is the service provider who is obliged to undertake reasonable adjustment to

Figure 8.2 The Grade 2 listed URS Building on the Whiteknights Campus. (University of Reading) – Extract from the official listing: *Architectural interest: the expressive use of structure to enclose space, which references traditional Japanese construction, and the playful exaggeration of the post and lintel joints, give the building drama, wit and virtuosity.* This is an example to post-war modern architecture which is being increasingly listed.[6]

facilitate access, it is the registered building owner who is responsible for ensuring that a listed building is not altered without the relevant consent. Furthermore, if works are undertaken without permission, irrespective if these are undertaken by the service provider who may be a tenant, it is the owner who is liable to prosecution. In most instances where a commercial listed property is not owner occupied but tenanted, there is an appropriate clause in the lease contract obliging the tenant to apply for a licence to alter the property. This is something that may be applicable for the provision of implementing access solutions but as part of any proposal to undertake alteration to a listed premises, this will also need official approval under the listing.

Accordingly, there will be a need to seek planning approval be means of listed building consent which is administered by the relevant local planning authority and will require the necessary discussion with the appointed conservation officer. There is a need in the listed building consent to make reference to the importance of the architectural and historical importance of the building elements to detail how the proposed alteration with affect these. While it has been discussed the notion of 'reasonable' adjustment in the context of that favouring service providers and also the need to respect the 'reasonable' adjustment of those experiencing disability, there should also be some consideration on what is 'reasonable' in the context of alteration to the historic building fabric. While there may be a temptation to use that argument that 'when it's gone, it's gone' to prevent alteration for accessibility, this no doubt would favour preservation and, in many ways, align with the approach of SPAB. Accordingly, this appears to create inaccessible 'museum pieces' of buildings which cannot truly be enjoyed by all, this certainly does not reflect the approach of Historic England and the design to enhance the significance of heritage buildings. It should be noted that some temporary access solutions may also require listed building consent but such items as portable ramps that are folded away and installed as when needed does not need permission.

The process of audit associated with assessing accessibility is binary, as this will always compare the situation of an existing, heritage building with modern legislation or best practice, there no doubt will be a need to compromise. Added to this is the notion of reasonable adjustment which is subjective in the context of the disabled user, service provider and heritage asset. Therefore, it can be stated that the retrospective design solutions for the provision of accessibility adopt the principles of honesty, sympathy or a combination of both.

The concept of honesty in respect of undertaking reasonable adjustment to facilitate access develops largely compliant design solutions executed with modern generic materials. This is typically exemplified by the use of lightweight metallic ramps and supports effectively bolted on to the external entrance of a building. This is honest in its recognition of an access problem met with an access solution that does not appear to make any attempt to blend in with the existing heritage. Honest solutions can also have aesthetic purpose, typically with the placement of an external ramp, and executing this is completely different materials recognising that there is a fundamental difference between old and new. In complete contrast is the implementation of an access solution, which is in keeping or sympathetic with the existing heritage asset. This may engender a compromise in some of the materials or detailing to ensure uniformity with the old and new. While the sympathetic option may result in an access solution which appears to blend the different age categories, it is important to minimise solutions which deviate away from best practice as this may affect total inclusivity for the sake of aesthetics.

Retrofitting design solutions which require modern technology such a lifts with invariably introduce something that simply did not exist at the time of the original building construction. While it is possible to introduce finishes to the lift cars as an alternative to typically modern brushed stainless steel wall linings, it is still necessary to consider colour, contrast, lighting and shadow. Furthermore, any attempts to change the formatting of control buttons or signage away from the norm would potentially cause more problems.

Ultimately, and in the event that an access audit of a commercial listed premises identifies and makes recommendations concerning reasonable adjustment, this will need to be applied through the relevant legal prescriptions. In the event that listed building consent is not granted, an appeal cannot be made on the grounds that failure to make reasonable adjustment contravenes the Equality Act 2010. This was discussed over a decade ago in the Houses of Parliament, and the published parliamentary briefing paper concluded that:

> *The Disability Discrimination Act, in common with the Sex Discrimination Act 1975 and the Race Relations Act 1976, cannot require anything to be done that would contravene another piece of legislation. Where a service provider must get statutory consent to a particular alteration, including listed building or scheduled monument consent, and that consent is not given, the Disability Discrimination Act will not have been contravened. However, a service provider would still need to take whatever other steps under the Act were reasonable to provide the service.*[7]

Added to this suggestion that the making of a reasonable adjustment must not contravene fire legislation or contribute to a health and safety risk. It would appear that the obligation placed on a building occupier to carry out a fire risk assessment or health and safety audit should be inclusive in the provision of a safe evacuation for all. However, if the proposed placement of an accessible feature introduces a risk, the Equality Act should be overridden by the relevant health and safety. An example of this may be the siting of an accessible toilet or shower cubicle to an office where the access door opens outward into an emergency escape route.

Historic England has produced an excellent guidance document concerning the process and rationale for providing access solutions to listed buildings.[8] This is a publicly accessible and free document. It is also possible to research and review examples of access proposals through other online resource. Of particular note was a consultation between English Heritage, York Museums Trust and City of York Council to make York's Clifford Tower fully accessible for wheel chairs.

To place this in context, the tower is a Grade 1 listed building and was initially constructed by William the Conqueror in 1068 and sits at the top of a large earth mound in the centre of York[9] (Figs 8.3 and 8.4).

Access to the tower is afforded by a steep set of stone steps to the southeast side of the grassy bank. The subsequent summary of the consultation available via English Heritage[10] and considered a number of options to facilitate the 10 m of vertical travelling distance, these included placing a spiral ramp around the mound which would amount to approximately 270 m of inclined pathway. This and the construction of an

Figures 8.3 and 8.4 **Cliffords Tower York** (photos courtesy of Lorraine Farrelly).

Figures 8.5–8.8 **Above: Image of the Tower of London and the key pages of the Historic Royal Palace's Access Guide which uses Red/Amber/Green coding to illustrate area of accessibility.**

external tower lift with bridge to the historic castle keep were discounted as it would significantly alter the visual concept of the tower which sits on a grassy hill above the city centre. A further design solution to tunnel into the mound and place an internal lift shaft within the tower was also rejected by virtue that it was not possible to facilitate an emergency escape route from the tower utilising the existing narrow stone spiral staircases.

There is documented evidence that post consultation, it was concluded that there was no viable solution to create step-free access to the tower. This may be considered relatively rare that any part of a historic building was deemed inaccessible to wheelchairs and further noted are other significant historic buildings which seek to provide the best possible access.

Noted are other examples of high profile heritage buildings that are open to the general public and recognise the access limitations. The Tower of London is a Grade 1 listed building and also originally built by William the Conqueror dating 1070. It is a landmark building which was never intended to be accessible, however, owned and operated by the Historic Royal Palaces (HRP), there has been an attempt to make reasonable adjustment. According to the access guide published by the HRP, there is a positive attempt to facilitate access with an honest recognition of the limitations imposed by the existing historic fabric.[11] The proactive approach to identifying and explaining what is accessible as well as the reasons why access cannot be permitted is an example of choosing opportunity over obligation in the provision of an inclusive environment. If this can be done for a building which pre-date accessibility legislation by approximately 900 years, there should be no reason why a similar approach cannot be adopted by relatively modern commercial properties as well as the providers of goods, services and employment (Figs 8.5–8.8).

The notion of reasonable adjustment to a historic building is restricted by the need to protect the heritage value. It has been considered the concept of removing or alter a barrier to access or even the possibility of avoiding this or providing an alternative means of access.

Having established the significance of the historic built environment in the terms of its historical and cultural importance as well as the positive contribution this makes both socially and commercially, it has been observed the challenges associated with providing access. As previously indicated throughout this text, new build or renovated commercial properties which complied with the relevant building regulations at the time are likely to comply with accessibility requirements. All other buildings and in particular those deemed to contribute to the historic built environment will require retrospective measures to accommodate reasonable adjustment. It is important to recognise that while there are some admirable attempts to adopt technical solutions as well as the associated compromises, it is still necessary to adopt a best practice approach.

Technical Solutions and Compromise – Parking

The obvious concern regarding accessible parking to the historic built environment is that many of the buildings were conceived at a time when there simply were no cars. Furthermore, a significant quantity of listed commercial properties is located in cities and town centres across the UK. Therefore, the very nature of these locations combined with pre-1920s construction means that many of these properties are unlikely to have their own individual parking allocation. Accordingly, parking is likely to be provided by street parking under the control of local authorities or purpose built single or multi-storey car parks operated by private parking providers or the local authority. The very nature of commercial car parks and parking buildings is such that the relevant service provider is obliged under The Equality Act 2010 to make appropriate reasonable adjustment to facilitate access. Existing car parking provision should be relatively straightforward to accommodate accessible parking spaces; these are typically wider than non-accessible spaces. Accordingly rearranging and remarking existing car parking spaces is a straight forward technical exercise but it may require converting three existing spaces into two accessible spaces.

One of the biggest concerns regarding accessible car parking spaces is the available numbers of spaces, and there is prescriptive guidance concerning how many accessible spaces should be allocated in a car park within British Standard BS8300-1:2018. As previously indicated, it is unlikely that there will be bespoke car parking allocated to historic buildings due to the very nature of these pre-dating this mode of transport. However, parking may be present to historic buildings operated as tourist attractions; this may include the placement of parking spaces on land adjacent to or within the grounds of the property. It is not sufficient for there to be token parking allocation for those with a

Figure 8.9 and 8.10 Accessible Parking to a Grade 2 listed railway station (1856–58), despite this it should be noted the dropped kerb which is absent of tactile paving.

physical or mental impairment; if this is present then it should have all the required characteristics. This includes correct dimensions and clear widths as well as flat, level and smooth surfaces. Signage, lighting and the avoidance of earth, gravel or cobblestones should be considered with respect to the interface between the parking, access route and building entrance. While it is accepted that car parking within the boundaries of a listed building or located in a conservation area should be sympathetic in appearance with the existing historic fabric, it will be necessary to still use appropriate materials (Figs 8.9 and 8.10).

Access Routes

As previously discussed, the external access routes either from onsite parking or the building approach from private car parks or those operated by the local authority should take into consideration gradient, material type and ensure these are non-slip, firm, as well as durable materials. Where the external access route comprises the surrounding streets in a town or city location, these are under ownership of the local authority, and therefore, the building owner or service provider has little or no control over these. While many historic buildings on the high streets of town and city centres interface directly with the local authority streets, there are instances where there may be an access route connecting the local authority street with the building or access routes from onsite parking to the main building entrance.

To facilitate an inclusive environment these, need to be wide enough for two wheelchairs to be able to pass each other, not too steep in gradient as well as being firm, durable and non-slip. There appears a tendency with historic buildings to execute car parks and access routes to be finished with loose gravel or stone chippings; these are not suitable for wheelchairs and can also be restrictive to those with other walking or mobility aids. Access routes also need to be well lit and the principal entrance to the building appropriately signposted (Fig 8.11).

Figure 8.11 Access route to a Grade I listed building from onsite car parking with step free access provided by an alternative access route. The external access steps adopt an 'honesty' over 'sympathy' approach and have been executed in concrete with galvanized steel handrails.

External Vertical Circulation

Access routes from adjacent streets or onsite parking may have to accommodate changes in level. While modern buildings are encouraged in their design to incorporate the topography to provide access routes to the main entrance, this is nearly always something that has to be implemented retrospectively with heritage buildings. Changes in level can be accommodated with the provision of access ramps or external platform lifts. When considering the implementation of these, it is still necessary to foresee where possible the specific criteria outlined in the guidance provided by BS8300-1:2018. This will include the relevant gradient of any ramp, the width of this, the positioning and specification of handrails as well as the use of tactile paving (Figs 8.12 and 8.13).

Main Entrance

It can be argued that alteration to main entrance of a heritage asset has the potential to cause significant harm to the building fabric. In essence, the principal entry point to any building is a magnet to activity as well forming the conduit in which generations of previous occupants or users have brought life into the premises. The most obvious concern regarding entry is the requirement for this to be physically wide enough to accommodate wheelchair access. In line with the principles of eliminating an obstacle the widening or adjusting of the entrance may be seen as a viable option. However, there are numerous examples where it is possible to create an alternative entry. One of the key

Figure 8.12 Above: External vertical circulation to a 15th century Grade I listed building which appears to choose 'sympathy' over 'honesty' in its design. The use of matching materials and aesthetics appear in keeping with the existing building although some of the specific detailing does not match best practice criteria.

Figures 8.13 and 8.14 Above (left): Access steps to an entry/exit door of a 16th century, Grade I listed, multifunctional building, the external vertical circulation restrictions are navigated with a concealed platform lift. With modern technology it is an honest approach to reasonable adjustment.

considerations with using effectively a secondary entrance for those with a disability is that it is possible to draw a parallel perspective in the sense that such action is treating those with a disability as 'second class citizens', which would go against the very principles of the Equality Act 2010 which prevents discrimination. As part of the access statement associated with any decision to provide a 'secondary' principal entrance is whether this actually facilitates access to the same point or service provision with relatively little extra deviation (Figs 8.15–8.17).

Having established the citing of a suitable principal entrance and the potential risk that this poses to the heritage value of a listed building, there is certainly a need to consider making this a power-assisted facility. Research undertaken has identified that for those who recognise as having a disability categorised to affect their mobility, cite the provision of a power assisted door as an essential (not reasonable) adjustment. There is no doubt that a power-assisted entry door constitutes an 'honest' approach to this; it is wholly inappropriate to pretend that such technology existed when the original building was constructed. Despite this, there are ways to adopt the technology and increasingly there may be the possibility to retrofit existing historic fixtures and fittings to integrate a means of mechanism while retaining the visual appearance associated with the historic fabric (Figs 8.18 and 8.19).

Figures 8.15–8.17 A Grade 2 listed bank building dated 1876 still in use as a bank, despite utilising an alternative entrance to navigate the external steps, the clarity of the access route as well as the presence of vegetation presents a hazard or obstacle.

Figures 8.18 and 8.19 Reading Town Hall, library and museum – Grade 2* built between 1871 and 1897. Renovated to include a power-assisted, fully glazed, sliding entrance door to facilitate access.

Other key considerations regarding the retro fitting for accessibility and the principal entrance includes signage, lighting, and, if possible, the implementing of a weatherproof canopy or protect. However, the placing of the latter is likely to have a negative impact on visual appearance of the heritage asset and in particular the façade.

Horizontal Circulation

As established, large quantity of historic commercial property is located on the 'high street' and one of the defining characteristics of pre-1920s high street properties is that these are typically built using loadbearing walls. These can comprise the external walls of the building envelope with floor and roof joists built into these, quite often there are internal walls which contribute to the internal circulation areas. It is not uncommon for the internal walls also to be loadbearing, as a consequence, this may restrict significantly the internal circulation areas. Any subsequent plan to remove part or all of the internal walls to facilitate access will require listed building consent but may also compromise the structural stability of the building (Figs 8.20–8.22).

The desirable clear width of internal corridors according to the best practice guidance is 1.8 m, and in instances where the physical structure or placement of loadbearing internal walls a width of 1.2 m can be permitted with the possibility to provide passing places.

(caption on next page)

Figures 8.20–8.22   **Above (top):** A commercial retail unit which appears to date the early 20th century, further examination and research confirms this to be a 16th C Grade 2 building. Noted is the narrow access route between the two halves of the property. Altering this would remove the structural timber frame.

Vertical Circulation

As alluded to on a number of occasions in this textbook, vertical circulation is probably one of the single biggest challenges to overcome for those with a physical impairment. In context, the notion of mechanical vertical transport was not developed until the 19th century. These would have been generally only for buildings of the highest status or those with a bespoke requirement to transport people, equipment or goods. They certainly would not be seen as a requirement for the listed buildings associated with low rise 'high street' commercial property. Indeed, even in the intervening 200-year period since the initial commercial invention of lifts, the majority of low rise, historic buildings are not equipped with one.

There are essentially two key reasons why there is still not widespread adoption of lift technology to low-rise commercial buildings as a general concept. The first is that is a high investment cost, and most organisations undertake a cost-benefit exercise to see if this is viable. Alongside the initial outlay to install a lift are the costs associated with running and regular maintenance. The second reason is that there is a technical requirement for installing vertical transport between different floors. Where this is going to be a 'traditional' lift, there effectively needs to be a hole cut between the different vertical levels and some kinds of enclosure or lift shaft created. The equipment itself comprises a number of different technologies as there are some lifts that pull the weight from above, and these are known as traction lifts. These types of lifts require a lift machine room to be positioned above the highest level served by the lift car. The other type of lift is a hydraulic lift, and this technology pushes the lift from below and does not need a lift machine room.

Other lifting devices include platform lifts or stair lifts, which can be used typically to negotiate one level or less of vertical travelling distance. These are quite common to low-rise commercial properties and generally do not require the placement of a lift shaft (Figs 8.23–8.25).

The notion of reasonable adjustment has been discussed on numerous occasions, and while there is presently a focus identifying reasonable adjustment from a service provider or employers' perspective, this should focus more on those with disability. What is reasonable to an employer or service provider may be significantly different from the to the perception of reasonable adjustment by the disabled user or employee. However, in the case of retrofitting of a lift installation to a historic building, there is a case to strike a balance between the two differing opinions on reasonable adjustment. Firstly, it may be technically impossible to install a lift as seen with the Clifford's Tower proposal, and in this instance, there was simply no way of providing any step-free access. However, this building and some of the other historic castles, palaces or monuments are quite rare examples of commercial listed buildings. Most commercial listed properties are in the retail and leisure sectors providing shopping, restaurants, pubs, cafes, hotels, and guest houses.

Figure 8.23–8.25 Top left: A decommissioned historic lift serving four levels to a Grade 2 listed building dating 1911. Bottom left: A retro fitted platform lift to a Gade 2 listed building dating from the late 18th century with a current retail use. Right: A retrofitted lift to a 16th century, Grade 1 listed building.

The concept of being able to remove, alter, avoid or deliver alternative access to goods, services, and employment can simply be by bringing the service to the disabled user as opposed to facilitating access to the service. For example, it may be possible to provide the same standard or food and dining experience on the ground floor of a two storey restaurant. This seems to be an acceptable alternative to installing a lift unless of course the restaurant markets the dining experience with say 'excellent panoramic views' from the first floor restaurant. With the example of a retail unit illustrated in the photo above, it is simply not possible to provide all of the clothing from the lower ground floor to the ground floor for a disabled customer. This is not uncommon, and in most unisex clothes shops, it is not uncommon for male, female and children's clothing to be situated on different floors. In this instance (as illustrated), a platform lift has been installed.

From a service provider or employers' perspective, the installing of a lift is a highly costly exercise and in line with guidance published on the application of The Equality Act 2010, 'cost' is considered to be one of the reasons why retrofitting might be unreasonable. In the event of a legal claim for discrimination based upon the unavailability of goods and services through the inability to navigate a vertical-level change, there may be a forensic examination of available finance. The relative lack of successful claims for discrimination for those with a disability against service providers is testimony to perhaps how difficult it is to prove reasonable adjustment. Furthermore, as identified in the Paulley V First Group PLC case, there is criticism that the legislation has no 'teeth' as ultimately and despite the outcome, little has been done to force change.

Providing access to goods, service, and employment should not be a race to the bottom and, as previously established, the stick as opposed to the carrot does not appear to have worked. There is a notion that adopting opportunity over obligation can have a more positive outcome for both the service provider, employer, and those with a physical or mental impairment. Another example of this may be with the provision of an accessible bedroom to a small independent hotel or guest house, which is listed (or in general terms not listed). There appears little justification in both technically and financially installing a lift based upon the cost benefit of this, irrespective of whether this is technically possible. However, by adapting the property to provide an accessible bedroom on the ground floor, this may suffice as a reasonable adjustment.

Signage and Communication

The provision of suitable signage within the historic built environment should be relatively straight forward to implement retrospectively. There is clear guidance on the type of signage, lettering specification, as well as the use of standardised pictograms. The retrofitting of signage within an existing heritage building should have minimal impact on the existing building fabric and structure, in theory, any works required to fix, place or operate signage should be fully reversible if it is deemed necessary to put back the existing to meet with the requirements of the listing. Furthermore, there appears no conceivable reason why it is not possible to use tactile signage within a heritage building.

One of the most effective ways to facilitate communication for those with hearing loss of deafness is by installing hearing loops to the reception area of office or hotel accommodation; this is also beneficial to retail properties or cafes, pubs and restaurants to the area where counter service is available. The integrating of hearing loops can be undertaken with the operational fit out of the spaces, where this involves cutting into the existing historic building fabric, there will be a need to seek the appropriate listed building consent.

Sanitary Rooms and Facilities

The provision of sanitary rooms should be relatively straight forward to adopt for accessibility on the basis that where there are existing abled bodied toilets it is prudent to use the existing water supply and evacuation pipework. The principle only works if the sanitary rooms are located on the ground floor, or if there are appropriate facilities to navigate the required vertical circulation for sanitary rooms located on different floor levels. It is quite common for toilets in restaurants and to be located on the first-floor level or in the basement. In most normal circumstances, this is done to free up space on the ground floor to operate the prime operation of the service provider; accordingly, this can introduce the need for a supplementary assessable toilet cubicle to be installed. One

Figures 8.26 and 8.27 A bank within a consevation area converted into a coffee shop with an accessible, mixed gender toilet cubicle constructed internally behind non loadbearing stud walls with a bookcase finish which is fully reversable alteration.

viable option for relatively small properties is to install a mixed gender and accessible cubicle (Figs 8.26 and 8.27).

The provision of other generic facilities such as accessible bedrooms or waiting areas, counters and reception desks should be relatively straight forward to adopt for heritage assets. As previously discussed, the notion of providing an accessible bedroom for a relatively small hotel or guest house does not always have to mean that there needs to be a lift installed. However, it is accepted that concerning town centre guest houses which may comprise narrow, multi-storied town houses converted accordingly, the nature of these properties may mean that there are limited ground floor rooms to reconfigure for accessibility. In line with some of the complex historic case studies opened up for examination, there may be a point at which it has to be recognised that it is simply not possible to provide such a service. By adopting the philosophy of seeking opportunity over obligation regarding accessibility, there should be a recognition that access is provided by those that want to do it, not because they have to. There is widespread acceptance that there is commercial value in providing an inclusive environment,[12] and once the cost-benefit of implementing accessibility is established, it has the potential to be a win/win for commerce and society.

Perhaps more challenging is the adoption of inclusive design to those heritage assets which provide audience and spectator facilities. One such venue is The Old Vic Theatre in London, which has Grade 2* listed status, the service provider openly advertises the fact that the venue is accessible.[13] Not only are there a number of wheelchair spaces and accompanying spaces for companions or carers, but there appears an all-inclusive approach to theatre performances including[14]:

- audio description and captioned performance;
- touch tours;
- braille programmes; and
- infra-red audio enhancement system.

In summary, it appears too easy to avoid undertaking retrospective measures to make heritage assets accessible on the basis that they are listed. It is a similar argument with the need to improve energy efficiency to old buildings, and if these are listed, they do not have to conform to the requirement for an Energy Performance Certificate. Regardless of this, the occupier or users still pays for the consumption of energy associated with heating and lighting. Improving energy efficiency will ultimately have a positive environmental impact, but it also reduces the occupational costs. This too is a win / win for the user and wider society.

There will no doubt be compromises that have to be made to facilitate access, and this is one area of inclusivity where is it necessary to think outside of the box. The access audit process as discussed in Chapter 6 is binary, it will give an immediate sense of whether the commercial building, goods, services or employment is accessible. However, it is the rational discussion and preparation of design solutions, which is a measure of the will, desire and creativity of an organisation to choose opportunity over obligation, which is in part the driver. There is a formal application process and legal conformities that need to be addressed; however, there are numerous worked examples of how access can be achieved. There may be a realisation that with some properties access may simply not be possible to all or parts of this; accordingly, there is a need to undertake a forensic due diligence. An evidence-based, well-researched access statement goes a long way to respecting the process and can see off potential legal challenges. Aside to this, improvement in technology and in particular virtual reality can be beneficial for some user experience activities such as museums or galleries.

Having established the history of disability and definitions as well as examining the notion of commercial property and ways to deliver access to all sectors including the historic built environment, it necessary to begin to visualise access solutions. Chapter 9 seeks to draw much of the discussion together and apply the theory with an examination of the practical application of design solutions.

Notes

1 Introduction to the Heritage Protection Guide: Heritage Conservation Defined | Historic England
2 About the Society for the Protection of Ancient Buildings (SPAB)
3 SPAB Approach
4 What are Listed Buildings? How England's historic buildings are protected | Historic England
5 Microsoft Word – Revised Principles of Selection 2018 (publishing.service.gov.uk)
6 URS Building, including the paved surface of Chancellors Way and raised edges of the ornamental pool, University of Reading, Earley – 1435127 | Historic England

7 SN03007.pdf (parliament.uk)
8 *Easy Access to Historic Buildings (historicengland.org.uk)
9 History of Clifford's Tower | English Heritage (english-heritage.org.uk)
10 improving-access-to-cliffords-tower.pdf (english-heritage.org.uk)
11 tower-access-guide-2018.pdf (hrp.org.uk)
12 Public-and-Commercial-Attitudes-to-Disability-in-the-BE.pdf (ucem.ac.uk)
13 OV Access | Old Vic Theatre
14 Making Theatre Accessible | Old Vic Theatre

References

Department for Digital, Culture, Media & Sport. (2018). *Principles of selection for listed buildings.* [online] Available at: https://assets.publishing.service.gov.uk/government/uploads/system/uploads/attachment_data/file/757054/Revised_Principles_of_Selection_2018.pdf [Accessed 14 January 2024].

English Heritage. (n.d). *History of Clifford's Tower.* [online] Available at: https://www.english-heritage.org.uk/visit/places/cliffords-tower-york/history-and-stories/history/ [Accessed 14 January 2024].

English Heritage. (n.d). *Improving access to Clifford's Tower and the wider York Castle site.* [online] Available at: https://www.english-heritage.org.uk/siteassets/home/visit/places-to-visit/cliffords-tower/gallery/improving-access-to-cliffords-tower.pdf [Accessed 14 January 2024].

Historic England. (n.d). *Listed buildings.* [online] Available at: https://historicengland.org.uk/listing/what-is-designation/listed-buildings/ [Accessed 14 January 2024].

Historic England. (n.d). *Planning.* [online] Available at: https://historicengland.org.uk/advice/planning/ [Accessed 14 January 2024].

Historic England. (n.d). *URS building, including the paved surfaces of Chancellors Way and raised edges of the ornamental pool, University of Reading.* [online] Available at: https://historicengland.org.uk/listing/the-list/list-entry/1435127?section=official-list-entry [Accessed 14 January 2024].

Historic England. (2012). *Easy access to historic buildings.* [online] Available at: https://historicengland.org.uk/images-books/publications/easy-access-to-historic-buildings/heag010-easy-access-to-historic-buildings/ [Accessed 14 January 2024].

Historic Royal Palaces. (n.d). *Access guide: Visiting the Tower of London.* [online] Available at: https://www.hrp.org.uk/media/1571/tower-access-guide-2018.pdf [Accessed 14 January 2024].

SPAB. (n.d). *SPAB – About us.* [online] Available at: https://www.spab.org.uk/about-us [Accessed 14 January. 2024].

SPAB. (n.d). *The SPAB approach.* [online] Available at: https://www.spab.org.uk/campaigning/spab-approach [Accessed 14 January 2024].

Tagg. (2020). *Public and commercial attitudes to disability in the built environment.* [online] Available at: https://www.ucem.ac.uk/wp-content/uploads/2020/10/Public-and-Commercial-Attitudes-to-Disability-in-the-BE.pdf [Accessed 14 January 2024].

The Old Vic. (n.d). *Making theatre accessible.* [online] Available at: https://www.oldvictheatre.com/making-theatre-accessible/ [Accessed 14 January 2024].

The Old Vic. (n.d). *OV Access.* [online] Available at: https://www.oldvictheatre.com/visit/ov-access/ [Accessed 14 January 2024].

Wilson. (2013). *Disabled access to historic and listed buildings.* [online] Available at: https://researchbriefings.files.parliament.uk/documents/SN03007/SN03007.pdf [Accessed 14 January 2024].

Chapter 9

Design Solutions and Creating an Inclusive Environment

Introduction

When considering the application of measures to deliver an inclusive environment, it is necessary to establish if the commercial building that houses the goods or service provider is an existing building, renovation or new build project. This will have implications on whether alterations and changes are a genuine compromise to navigate historical access issues or can be adopted into the design scheme from the offset.

Specific research into the access of commercial buildings, goods, services and employment has established a priority rating on the experiential access needs of those with a disability.[1] While it has been discussed that goods and service providers as well as employers are obliged to undertake reasonable adjustment to facilitate access, this appears based on the legal minimum requirement. Furthermore, it is also based on what is reasonable to the service provider or employer and not the disabled user. Therefore, the research provided respondents with access criteria based on the adoption of *Best Practice* principles, giving them the following options concerning their access needs responding to all facets of access:

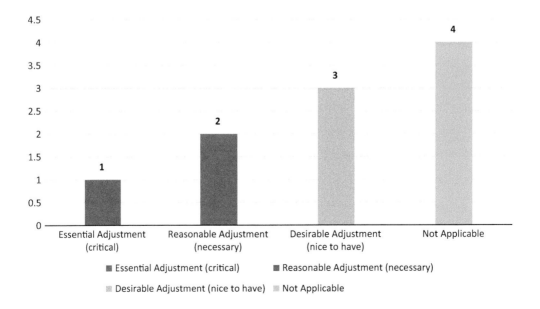

DOI: 10.1201/9781003239901-9

There are a number of ways to analyse the data in the study, and it is important to contextualise the priorities of those with a physical or mental impairment and their specific experiential needs for access. One simple method was to create an average score, which gives an indicative trend, looking deeper into the data and dependent upon the number of respondents per disability or category of disability; it was possible to refine and establish the individual disability specific needs.

What is evident from this research is the notion that most of the best practice guidelines for accessibility are deemed an **essential adjustment,** which is critical in nature. This appears to represent a significant gap in perception on access adjustments deemed 'reasonable' by service providers and building owners. Perception gaps are not uncommon between those with a disability and those without, as well as a commercial perception gap detailed in Chapter 7.

To understand and contextualise the implementation of best practice, it is necessary to visualise the current uptake of largely legal minimum measures to deliver an inclusive environment. There is a need to consider the requirements for legal minimum compliance and detail the necessary upgrade to ensure this meets with best practice guidance. In many cases, the upgrade from the legal minimum to best practice appears relatively straightforward, although it is accepted that this comes at an additional cost, but in some cases, the benefit outweighs this. Concurrently previous research has indicated that there is commercial value in providing an inclusive environment.

In line with the legal prescription, best practice guidance and the access audit process, it is necessary to consider the following accessible features when determining the implementation of solutions to achieve accessibility:

- parking;
- access routes;
- external vertical circulation;
- building entrance;
- horizontal circulation;
- vertical circulation;
- signage and communication; and
- facilities and sanitary rooms.

The following fact files are a summary analysis of solutions to permit an inclusive environment based upon the publicly available minimum legal standards. These are stipulated in the Building Regulations, Approved Document Part M Volume 2 – *Access to and use of buildings*. The discussion in this chapter is a summary of the key points; for further detailed clarification, it is necessary to cross reference the summary against the published legal framework:

- Approved_Document_M_vol_2.pdf (publishing.service.gov.uk)

Attempts to provide an inclusive environment building upon the legal minimum requirements should endeavour to adopt *Best Practice*, according the fact files also detail a summary of requirements contained within British Standard 8300, which seeks to deliver best practice in this matter with the following documents:

- BS 8300–1:2018 Design of an accessible and inclusive built environment. External environment – code of practice and
- BS 8300–2:2018 Design of an accessible and inclusive built environment. Buildings – code of practice.

The fact files summarise and interpret to some of the key requirements for best practice, but to ascertain full and comprehensive analysis of these requirements, it is critical to consult the exact information contained within these publications. These can be found by accessing the information from the BSI Group

- Standards, Training, Testing, Assessment and Certification | BSI (bsigroup.com)

While the fact files detail many of the key considerations for access, these should not be used as an explicit reference for the design and audit for accessibility within the built environment. As previously stipulated, it is necessary to refer explicitly to the existing relevant building regulations and British Standards accordingly.

Parking

FACT FILE: Designated on Street (Parallel) Parking

Figure 9.1   *Figure 9.2*

The minimum legal prescription concerning parking does not specifically cover designated onstreet car parking but illustrates standard parking bays that are typically located with purpose-built parking areas. However, this is addressed with the design advice for best practice, which states the following:

- the space dimensions should be 6.600 m long and 3.600 m wide;
- where space permits the provision of one large space this should be 8.000 m long and 4.800 m wide;
- there should be in place methods to prevent misuse by non-disabled users; and
- there should be a dropped kerb and tactile paving.

The obvious constraint of installing allocated accessible on street car parking spaces is that this may be limited by the width of the existing road. The example above makes an attempt to widen the space to facilitate safe vehicle exit but it does this by firstly reducing the previous car space. There is also no dropped kerb or tactile paving to the pavement, which does not facilitate step-free access. Furthermore, the overall markings are poor, and it is necessary to recognise the limited road width which constraints this retrospective placement of the accessible parking space.

FACT FILE: Parking Spaces

Figure 9.3

Figure 9.4

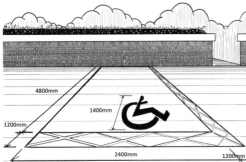

Figure 9.5 Design criteria (legal minimum).

Fig 9.3 New accessible off-street car parking bays located at a railway station appear largely compliant with the legal minimum requirement. The dimensions are compliant as is the flat, level and non-slip nature of the surface. Also present is a lowered (dropped kerb) and lowered adjacent section of paving with tactile (blistered paving). The presence of the raised planting boxes could potentially present an accessibility distraction, but these are outside of the zone of access. Fig 9.4: Worn finishes to basic accessible parking spaces emphasise the need to maintain spaces and markings.

To ensure that this complies with best practice guidance concerning accessibility, it is necessary to ensure the following extra requirements:

- provision of dedicated accessible parking spaces for employees;
- between 5% and 6% of overall car parking provision should be accessible which would represent in excess of 30 spaces for the whole establishment above;
- between 4% and 5% of overall car parking provision should contain enlarged spaces (3.6 m x 6 m) which would represent in excess of 20 spaces for the whole establishment above;
- appropriate signage is needed; and
- ancillary parking requirements such as the design criteria for barriers and ticket machines.

Research has established the following order of priority regarding car parking spaces: 1. The presence of dropped kerbs (1.27) / 2. Firm, durable non-slip surfaces (1.32) / 3. Clearly signposted and marked spaces (1.33) / 4. Level surface (1.37) / 5. Appropriate car parking space size (1.40). All of these are essential adjustments to those with disability.

FACT FILE: Setting Down Point

Figure 9.6

The legal minimum concerning a setting down point stipulates that this should be located on firm, level ground and a close as practicable to the principal or alternative entrance with the surface level with the adjacent carriageway. The setting down point should be clearly signposted and permit users to walk or travel by wheelchair to the entrance. The example above requires better positioning, signposting and the markings are also faded.

To convert this into best practice, it is necessary to consider the following regarding the setting down point:

- close to accessible building entrance;
- clearly indicated;
- setting down point additional to designated accessible parking places and taxi drop off point;
- if feasible short-term waiting provision should be facilitated for drivers waiting & picking up;
- setting down point should have minimum dimensions of 9 m x 3.6 m;
- where feasible the setting down point should be covered and at a minimum height of 2.6 m;
- dropped kerbs should be provided (where applicable); and
- the surface of access routes adjacent to the setting down point should allow for convenient transfer to and from a wheelchair.

Concerning the provision of both car parking and setting down points for accessing commercial buildings, goods and services, approximately 50% of disabled users deem this an essential requirement. Compared with the provision of dropped kerbs for car parking space (1.27), setting down points are the lowest access priority (1.73), but still essential.

FACT FILE: Ticket Machines & Barriers

Figure 9.7

The legal minimum for accessible ticket machines stipulates that these should be placed adjacent to the accessible parking spaces. These can be fixed to a plinth that does not project beyond the front of the machine with controls located between 750 mm and 1200 mm above the ground floor. The example above is not fully compliant as the meter is offset within the turning space (indicated by the red outline), and the placement of a parking bollard prevents access; furthermore, the height of the controls is outside of the dimensions specified.

Best practice compliance for parking meters includes the following:

- easily identifiable and contrast against the background;
- colour contrast of the buttons / controls against their backgrounds;
- provision of an alternative ticketing arrangements in the event of a failure;
- clear wheelchair turning space of 1850 mm×2100 mm in front of the machine; and
- the manoeuvring space should be level and free from obstruction.

Access Routes

FACT FILE: Width / Gradient / Materials

Figure 9.8

Figure 9.9

The legal minimum stipulates that an access route from the boundary or parking to the building should be level, 1500 mm wide with 1800 mm wide passing places 2000 mm in length. The approach should be no steeper than 1:60 across the length or 1:20 with landings and a cross fall of no more than 1:40. The images above illustrate the access route from a car park (top left), which has three steps leading up to the principal entrance. A signposted alternative, step-free level (1:60 or less) approach to a building is provided, but this does not appear to meet the width requirements.

The surface should be durable, firm and non-slip with similar frictional characteristics between different materials (if used) and undulations of less than 3 mm. Joints in paving should be no more than 5 mm, and where exceeding this amount, they should be filled flush or no more than 5 mm deep and 10 mm wide, if these are recessed. There should be measures in place to separate pedestrians and vehicles (missing in the example) and the route well lit.

Best practice identifies:

- widths or access routes should be 2000 mm wide to accommodate mobility scooters;
- overhanging obstacles such a canopies, sculptures or vegetation should not be lower than 2.5 m or penetrate the access route more than 150 mm;
- cross falls should not be more than 1:50 and drainage should be installed with cross falls to prevent water build up;
- where upstands are present, these should not be more than 150 mm high providing a tap rail for cane users and wheelchair safeguard;
- patterns and very dark areas of colour should be avoided as these can appear as holes to those with partial sight loss and
- lighting should be a minimum of 100 lux at floor level.

The essential experiential priorities for those with disability concerning access route are 1. Gradient (1.29) / 2. Firm, durable and non-slip surface (1.32) / 3. Flush and level joints (1.35) / 4. Access routes should be well lit (1.42) / 5. Access routes should be clearly identifiable (1.43) / 6. There should be provision of landings to long access ramps and gradients (1.44).

FACT FILE: Materials – Tactile Paving

Figure 9.10

Figure 9.12

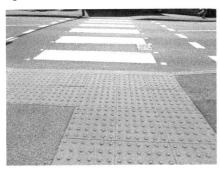

Figure 9.11

Tactile ('blister' Surface) Paving: This is a universally used concept and often formed using concrete paving slabs, which are precast to include 36 circular raised 'blisters'. These are dome-shaped and have a 25 mm diameter. These blisters serve as a hazard warning and can be felt under foot or detected with a long cane by those with sight loss or severe sight loss to indicate a crossing point of a carriageway. They are manufactured in four colours (buff / natural / charcoal / red). Buff and natural are the most widely used colours, and charcoal can be permitted for conservation areas.

Red (above right) is used to signify controlled crossing points, which could be in the form of a 'zebra crossing' or crossing points controlled by pedestrian lights.

They are incorporated with dropped kerbs and angled to ensure a smooth transition with the carriageway but are prone to colour fading with weathering, becoming soiled and dirt encrusted. The concrete can become eroded or worn, and they fracture if heavy vehicles drive on top of them (top left). In accordance with maintenance requirements, these should be periodically cleaned and replaced if they are cracked, dislodged or loose and present a trip hazard.

FACT FILE: Signage / Hazards / Barriers / Restrictions

Stainless steel protective bollards with yellow contrasting markings provide a separation from vehicles and pedestrians. They reduce in part the clear width of the path, but there appears sufficient passing place either side of the narrowing.

The bollards have a minimum height of 1.000 m and are aligned to ensure a uniformity in the obstacle that these present.

Figure 9.13

The descriptive in the legal minimum concerning access routes **does not explicitly detail how to address signage, hazard and barriers of restrictions**. Best practice addresses the issue of hazards and projections into access routes, it stipulates:

- street furniture (including lamp posts, litter bins and utilities cabins) should not be within the boundaries of access routes;
- where these are unavoidably present, they should colour contrast against the access route and background, and multiple obstructions should be in alignment to create uniformity;
- access routes should have a detectable raised edge along the duration for those with sight loss or severe sight loss;
- bollards and waste bins should be no less than 1.000 m high and contrast in colour;
- drainage channels and service outlets should not be within the boundary of access routes, and if these are evident, they should be non-slip, have slots at 90 degrees to the general direction of travel no more than 13 mm wide or 18 mm in diameter for circular drainage points;
- projections into an access route should not encroach by more than 100 mm, and projections of more than 100 mm into an access should not exceed 300 mm and have a leading edge at a height of 1200 mm above ground floor level;
- projections into an access route are permitted and unrestricted when this is above 2.500 m;
- projections encroaching into an access route beyond that permitted require the placement of hazard protection;
- access routes under stairs and ramps should be enclosed where the soffit is below 2.100 m, and accordingly, guard rails should be installed; and
- measures should be adopted to address the risk of falling where there exists a change of level between the access route and surrounding areas.

FACT FILE: Signage / Hazards / Barriers / Restrictions

Access control barriers at a railway station with a clearly accessible gate that has an intercom to call for assistance.

The materials used ensure there is sufficient transparency with the gates. The example illustrates a typical standard installation leading to the base of a staircase that is not step free. The same site has an alternative accessible lift with security barriers in front of the lift landing.

Figure 9.14

Examples of signage are explicitly covered later this chapter.

Barriers and gates should;

- be able to open in both directions, self-closing and operable with both hands;
- avoid pinch or twist operational mechanisms;
- have a clearance of 1.000 m; and
- have sufficient transparency or openness to be able to see users on the other side.

Concerning signage, best practice recommends the following:

- routes to parking, transport hubs, information centres and sanitary installations should be equipped with signage;
- signage should identify accessible and step-free routes;
- 'repeater' signage should be used on long distances;
- their shape, materials, colour and typeface should be the same in localised areas;
- signage locations should be visible from all directions, with appropriate best practice lighting levels;
- the placement of signage should not present as a hazard or obscure existing hazards or be concealed by shrubbery or vehicles;
- *You are here* orientation signage should be placed along access routes but not directly if these present an obstruction by those stopping to read these; and
- maps should include a north point, tactile embossing as well as audible information.

Well-lit (1.42) and clearly identifiable (1.43) access routes are two key adjustments regarding the experiential requirements for those with a disability accessing commercial buildings, goods and services.

External Vertical Circulation

FACT FILE: Steps and Stairs

Figure 9.15

Figure 9.16 **Corduroy surface and nosing.**

The legal minimum for external staircases stipulates:

- a level landing measuring 1200 mm x 1200 mm at the top and bottom of each flight;
- 800 mm wide 'corduroy' hazard surface should be installed 400 mm from the beginning and end of each flight;
- side access onto a landing should accommodate a 400 mm extension of the corduroy surface;
- no doors should open into a landing area;
- stair widths between enclosing walls or upstands should not be less than 1200 mm;
- single steps are not permitted;
- the number of steps is limited to 12 between landings where the width of each step is less than 350 mm or no more than 18 steps where the step width is 350 mm or more;
- a 55 mm contrasting strip should be evident to the nosing (vertically and horizontally);
- the projection of a step over the tread should be avoided but if present, this should be limited to 25 mm;
- the rise and going of the steps should be consistent in height for the duration of a flight and each riser should be between 150 mm and 170 mm;
- the going (width) of each step should be consistent and between 280 mm and 425 mm;
- risers are not open and continuous handrails are placed to each side of the steps / landings; and
- and the clear width between handrail should be not less than 1000 mm and not more than 1800 mm.

Best Practice adds to the legal minimum requirements with the following suggestions:

- the provision of weather protection of external stepped routes with risers between 150 mm and 180 mm with no more than 20 uniform risers;
- step widths between 300 mm and 450 mm;
- nosings should be contrasting in colour and slip resistant; and
- steps should have lighting levels of between 15 and 100 lux depending on location.

FACT FILE: Ramps

Figure 9.17

Figure 9.19

Figure 9.18

Figs 9.17–9.19 illustrate a range of attempts to provide ramped access with varying degrees of success. Fig 9.17 has a compliant gradient fitted with appropriate hand rails and a 100 mm kerb, but these have limited colour contrast with the ramp surface. The ramp in Fig 9.19 is too steep and is executed with glazed terracotta floor tiles that are not slip resistant. Fig 9.18 illustrates a ramp that appears too long with insufficient landings.

The legal minimum requirements concerning ramped access stipulates:

- ramps should be apparent or the approach to these are clearly signposted;
- the gradient of these should be between 1:12 for ramps of 2.000 m in length up to 1:20 for ramps 10.000 m in length with a maximum height of 500 mm;
- intermediate landings should be at least 1.500 long but alternatively measure 1.800 m x 1.800 m where these are passing places and should be installed where there are three or more sections of ramped access;
- a lift should be installed for wheelchair users where the vertical travelling distance is more than 2.000 m;
- landings should be free of any door swings or obstructions and have a gradient of 1:60 and max cross fall of 1:40;
- the ramp surface should be slip resistant and visually contrast with the landings with similar frictional characteristics evident between sections;
- handrails should be installed to both sides and a 100 mm contrasting kerb to the open side of the ramp; and
- alternative signposted steps are available where the height of the ramp is greater than 300 mm (typically two steps).

Best practice guidance places additional recommendations including:

- ramp width minimum 1.500 m;
- the provision of minimum lighting levels between 15 and 100 lux depending on the location of the ramp; and
- specific detail concerning the provision of handrails (see next fact file).

FACT FILE: Handrails

Figure 9.20

Figure 9.20a A plasticised coating to steel handrails ensures that these are less cold to touch than conventional painted steel. Noted also is the 100 mm kerb to either side of the ramp; however, the landings are not colour contrasted against the ramp surface.

Best practice and legal minimum requirements for handrails are the same as published for external steps and ramps. Concerning the legal minimum, *Building Regulations, approved document Part M* stipulates:

- the height of handrails should be between 900 mm and 1000 mm above the pitch line and between 900 mm and 1100 mm above the finished floor level of landings;
- where a second lowered rail is installed, this should be at 600 mm above the pitch line;
- handrails should extend 300 mm horizontally beyond the end of the ramp but do not project into and access route;
- these should be contrasting in colour but not highly reflective with them being slip resistant and not cold to touch;
- handrails are designed to terminate as to not pose a risk of catching clothing;
- they should protrude no more than 100 mm into the surface of the ramp and be either circular or non-circular in profile with a radius dimension of between 32 and 50 mm;
- handrails should be offset from the adjacent wall surface between 50 mm and 75 mm; and
- there should be at least 50 mm from the cranked support to the underside of the handrail.

Best practice guidance on handrails is similar to the legal minimum but adds the other following requirements:

- balustrades should have sufficient strength to withstand impact from mobility scooters, signage indicating a 4 mph should be evident;
- handrails should be installed to both sides of a ramp for unrestricted use; and
- they should be designed to withstand relevant loading.

FACT FILE: Lifting Appliances, Escalators and Moving Walkways

Figure 9.21   *Figure 9.22*

The provision of external lifting devices is not detailed in the legal minimum requirements, and these are not widely evident within the built environment because they are electro-mechanical devices that need to have adequate protection from the weather. This is detailed in the best practice guidance, and there is also a recognition that too cold temperatures can lead to condensation on metallic surface, along with solar gain, high temperatures and solar glare or high light levels, which may make controls difficult to read or use. Furthermore; where placed in areas of noise, it may be difficult to hear instructions or notifications.

The design criteria of external vertical transportation devices include:

- non-slip surface to the floor;
- adequate weather protection (roof) to the external lift well, entrance, controls & waiting space; and
- drainage facilities to ensure the evacuation of surface water from the lift pit or elsewhere (including the surface of platform lifts).

In most cases, platform lifts are the more common that escalators and moving walkways to facilitate external vertical circulation. Despite the rare use of external escalators, these are usually located to transport hubs, and where these are evident, users who are disabled are usually directed to lifts to accommodate vertical transport. Concerning accessibility, escalators are not the most appropriate form of vertical circulation and are clearly not accessible to wheelchair users or many ambulant disabled users.

Building Entrance

FACT FILE: Entrance – Placement, Identification and Access

Fig 9.23 is the principal accessible entrance to a commercial and the contrasting colours between to doors, entrance area and façade assist in the identification that this is the entrance, although it is not evidently signposted. The presence of a canopy provides weather protection; however, the steel struts supporting this constitute an access hazard.

Figure 9.23

The legal minimum requirements for accessible building entrances are that these are clearly identifiable, with appropriate signage from the site boundary or other entrances (if these are not accessible). The use of lighting or visual contrast should be used to make the accessible entrance easily identifiable from its surroundings, and any structural supports, porches or canopies should not pose a hazard to the visually impaired. The level landing area should be a minimum of 1500 mm x 1500 mm free from any door swings, and while door thresholds should be avoided, if these are present, they should be no higher than 15 mm with chamfered edges for any threshold greater that 5 mm.

Best practice largely aligns with the legal minimum but notes more emphasis or requirement on the following:

- Visual clarity: Entrance doors and fenestration (markings on glazing) should contrast against its surroundings and adopt appropriate lighting as well as being well lit. The use of mirrored glazing should be avoided and where possible it should be possible to see through the glazed doors into lobby or waiting areas.
- Weather protection: It should be foreseen to recess the building entrance or provide a weather proof canopy to the accessible entrance if the doors are not fully automated. The supports of any canopy should not present a hazard or restrict access.
- Threshold: The entrance door threshold should be level and a max height of 15 mm with any threshold over 5 mm having chamfered edges or pencil rounded.

Research data has identified the following ranking of entrance features for those with a physical or mental impairment: 1. Level entrance free from door swing (1.54), 2. Weather protected (1.85), 3. Identifiable / contrasting (1.95), 4. Clear view of internal waiting area (2.15) and 5. Canopy struts or obstructions should be avoided (2.28).

FACT FILE: Doors

Figure 9.24 Automatic entrance door with a flush threshold.

Figure 9.25 Access ramp at right angles without 1.5 m x 1.5 m landing and door latches that cannot be opened with a closed fist; it is not clear if the doors are automated.

The legal minimum prescription for entrance doors primarily addresses the necessary clear openings, and it should be noted that these are classified as either a new build installation or those to existing buildings. There are also nuances on whether the doors are accessible straight on or from the side with a 1500 mm x 1500 mm landing, as well as the requirement when there is insufficient turning space or if the doors are located near publicly accessible buildings. In essence, the minimum clear door with ranges from 750 mm for straight on existing buildings up to 1000 mm for new build publicly accessible buildings. For manually operated doors, there needs to be 300 mm clear space adjacent to the door on the handle side to allow wheelchair users to operate the handle. There are also clear limits on the force required to open doors, and where these have a self-closing mechanism, it is obliged to foresee a power-assisted door. Manually operated doors should have door latches that can be operated with a single closed fist, the door furniture should be contrasting and not cold to touch.

Power-assisted doors should encompass the following features:

• either manual control (push pad / card swipe / code access / remote control);
• automatic sensors with sufficient timing to open and delay closure to permit safe entry and exit;
• swing doors opening towards the operator have visual and audible warnings;
• they include a safety / stop sensor to prevent closing on the user;
• they default to manual operation in the event of mechanical or electrical failure;
• they do not project into an access route when open; and
• the controls are at a height of between 750 mm and 1000 mm above finished floor level and set back 140 mm from the leading edge of the door swing. Controls should be colour contrasting against their background.

Best practice aligns mostly with the legal minimum, but the minimum clear widths are 800–1000 mm. It recognises that automatic doors at the bottom of a ramp can present a risk and recommends guard rails if a door opens into access routes; it notes that if revolving doors are not accessible, a leafed door should be placed adjacent to these. The research has identified that door weight is the greatest essential adjustment for manual doors (1.27) and the provision of a safety stop mechanism for automatic doors (1.31).

FACT FILE: Glass Doors and Glazed Screens

Figure 9.26

Figure 9.27

The legal minimum requirement for glass doors and glazed screens identifies the risk of a person having a collision with these. Approved document Part M obliges the placement of manifestation (etching or marking on the glass) to provide a visual contrast across the whole glazed surface when viewed from both sides.

The design consideration is that whether the door is open or closed, it should be identifiable as should the leading / open edge of the door when open. The prescriptive detail as noted for the legal compliance defaults to section 7 of Approved Document Part K and is such that manifestation should be 150 mm in height and can comprise company logos etc. Alternatively, a rail can be placed on the glazed door or screen surface at heights between 850 mm and 1000 mm, as well as between 1400 mm and 1600 mm.

Best practice aligns with the legal minimum concerning fully glazed screens and additionally recommends that low-reflectance glass is used to reception and counter areas as this can negatively affect the ability to lip read.

Top (left) illustrates a power-assisted entry door to a commercial property that only has one low-level band of manifestation. While this does offer a degree of contrast and is visible from both sides, the manifestation to the fully glazed, power-assisted door (right) has strong contrast and is located at the two different band levels.

The placement of manifestation is designed to reduce the hazard risk associated with an individual colliding with full-length glazing; however, in the terms of a disability specificly identified by those with a physical of mental impairment, this was deemed a 'reasonable' and not essential adjustment (2.26).

FACT FILE: Building Lobbies Waiting Areas and Reception Desk

Figure 9.28 Insufficient knee space to desk.   *Figure 9.29* Sufficient knee space to desk.

The presence of a reception area will largely depend upon the commercial function of the building, and it should be noted that some buildings have an open or non-manned reception while others may have staff at a desk offering 24-hour coverage. Legal minimum requirements for reception areas include a general set of criteria that need to be adopted, where relevant. In essence, a building reception desk should be easily identifiable from the entrance but away from the doors to facilitate access. The approach should be able to accommodate wheelchair access and be direct as well as free from obstruction. The desk should be designed for use by those standing or sitting in wheelchairs; accordingly, there should be a raised and lowered section with the latter accommodating 700 mm free vertical knee space under the desk with a maximum counter height of 760 mm. To facilitate wheelchair access, there needs to be a free space or zone in front of the desk 1800 mm wide and 1200 mm deep; however, where there is no knee space under the desk, this zone needs to be 2200 mm wide and 1400 mm deep. The reception area should be equipped with a hearing-enhancement system, such as a hearing loop, and the floor of the reception area should slip resistant.

Best practice guidance covering reception areas additionally recommends:

- types and specification of non-slip floor materials and matting;
- placement of universal signage and pictograms; and
- provision of seating as an alternative to having to stand and wait in a queue.

Concerning the dimensions of basic lobbies and the space between the external building entrance doors (single leaf) leading to the following single leaf internal door, the legal minimum requirements stipulate the minimum distance between the protruding swing of the doors is 1570 mm. The default width of the lobby should be a 1200 mm or the width of the access doors plus an allowance of 300 mm to the opening edge of the door to permit wheelchair users to manoeuvre and operate any door handles. Glazing within lobbies should not create distracting reflections, and matting or changes in material should minimise frictional differences and not pose a trip hazard. Any projections within a lobby greater than 100 mm should be protected by a visually contrasting guard rail. According to the research data, the presence of non-slip flooring that does not impede wheelchairs or present a trip hazard is the most significant essential requirement for access (1.26) in a building reception; all other recommendations are found to be reasonable adjustment.

Horizontal Circulation

FACT FILE: Corridor Widths, Passing Places and Gradient

Figure 9.30 Internal corridor with contrasting wall and floor finishes, noted no contrast with the ceiling but the glazed screen to the lefthand side has manifestation.

Figure 9.31 Sloped horizontal circulation within a historic listed building, note contrasting walls and floor with handrails and non-slip carpet over ceramic floor tiles.

Corridor and passageways that facilitate internal circulation have a legal minimum requirement to be 1800 mm wide. There may be some obstructions, such as columns or radiators projecting into the space and reducing this width, and these should be fitted with a colour-contrasted guard rail. The legal minimum requirement also notes that where 1800 mm clear widths cannot be facilitated, then passing places of 1800 mm x 1800 mm should be present at regular intervals, and these can be in the form of corridor junctions.

Corridor floors and horizontal circulation should be level, but it is possible for these to slope with a gradient no steeper than 1:60; any horizontal circulation route greater than 1:20 should be designed as a ramp. Sloping floors between 1:20 and 1:60 should not rise more than 500 mm without a 1500 mm rest area; with sloping corridors, this should be across the whole width or the exposed edge should be colour contrasted and potentially protected with guarding.

Doors are not permitted to open into a major access route or means of escape unless these are recessed or the doors open into storage cupboards, shafts or cleaner's cupboards, etc., which are locked. Doors from accessible toilets can open into a main access or escape route if it maintains a clear width of 1800 mm. Accessible toilets with doors opening into corridors are permitted at the dead end. The placement of glazed screens to the internal corridor or circulation areas should include manifestation.

Best practice aligns mostly with the legal minimum requirements, although this does recognise that corridor widths must not be less than 1200 mm unless there are permanent obstructions over a short distance and, in the event of this, a width of 1000 mm is permitted. The research findings have ranked the following essential adjustment (<2.0) and reasonable adjustment (2.0–2.99) for horizontal circulation: 1. non-slip floor finishes (1.36), 2. Level and not sloping (1.51), 3. Sufficient width or presence of passing places (1.88), Free from projections (2.12) and plain non pattered floor finish (2.16)

FACT FILE: Floors, Walls, Ceilings and Lighting

Left: Fig 9.32 Internal horizontal circulation utilising contrasting floor and wall finishes, noted evidence of light reflectance on the non-slip floor finish. There is little visual contrast between the wall and ceiling finishes.

Figure 9.32

Figure 9.33 **Patterned floor finishes should be avoided.**

The minimum legal requirement concerning the finishes to the floors, walls, ceilings and lighting associated with horizontal circulation only addresses the floor finishes. It states that this should be slip resistant and avoid patterns that can appear as steps or level changes to those with sight loss. Best practice also stipulates that patterned flooring should be avoided, and in particular, reference is made to striped floor finishes that can be misleading for those with sight loss or neurological conditions. There is also specific reference in best practice to the effects of shiny floor surfaces that can cause glare.

Concerning the finishes to the walls, these should utilise materials that are not reflective, and this has been addressed when detailing the presence of glazed screens or partitions. The finishes to the walls should contrast with the floors and also with the ceiling. Best practice indicates that skirtings should be the same colour as the wall, but where this is the same colour as the floor, this should continue for no more than 100 mm. Internal doors and door frames should also be visually contrasted against the walls.

Best practice makes recommendations concerning internal lighting and specifically to areas of horizontal circulation where there is no natural light, artificial lighting should provide a level of 100 lux at floor level. Lighting should also be evenly diffused without glare or reflection, and this combination also requires an appropriate selection of floor materials. Research data has identified that contrasting floor, wall and ceiling finishes are considered to be a reasonable adjustment as is the presence of evenly diffused lighting that does not create glare. While these are noted to be reasonable adjustments, lighting is a priority at 2.07, almost equating to an essential adjustment. When considering specifically those with sight loss, lighting becomes an essential adjustment (1.67) while the remaining criteria remain relatively the same.

FACT FILE: Internal Doors and Escape Routes

Figure 9.34 Accessible escape route with signage, lighting and 'push bar' activation.

Figure 9.35 Magnetic hold backs securing escape doors in the open position.

The legal minimum prescription for internal access doors states the same clear widths as for the principal entrance, which is between 750 mm and 1000 mm depending on door position and if this is a new building. The door weight or force to open a door should not be more than between 22.5 N and 30 N depending upon the position of the door, with an initial 30 N maximum force required when the door is fully closed. A 300 mm clear space to the leading edge of the door is required where there is a door handle; any latch mechanism should be operable with a closed fist, and this and any door furniture should contrast with the door. Door frames should also contrast with the walls, and where the doors are held open, the leading edge should also contrast against the other door finish and surrounding surfaces. Doors or side panels wider than 450 mm should be fitted with vision panels that are positioned to the leading edge with visibility between 500 mm and 1500 mm above finished floor level, although this can be separated by the placement of an intermediate rail between 800 mm and 1150 mm; vision panels should include manifestation as detailed for glazed doors and screens. The legal minimum also notes that fire doors can be held back by electromagnets provided it release in the event of a fire or power failure. Fire doors in an escape route that have unequal door leaf widths should ensure that there is consistency in the placement of the same door width leaf to the same side of corridors.

Best practice aligns mostly with the legal minimum requirements but recommends marginal greater door clearance 800–1000 mm; it also addresses internal sliding doors. The levels of visual contrast are prescriptively specified in the best practice guidance concerning door frames, furniture and the leading edge. Guidance on the size and positioning, as well as the specification of vision panels, is given, including the minimum requirement for these to be 100 mm or wider. Glazed-vision panels should include manifestation to the glass.

The placement of door-closing mechanisms is detailed and described as being only applicable if absolutely necessary and these are typically required for fire doors. The force required to open the doors is the same as detailed in the legal minimum, and the research has identified this to be the highest priority (1.38) and 'essential', this along with colour contrast (1.83) and the ability to open handles with a closed fist (1.92).

Vertical Circulation

FACT FILE: Steps and Stairs

Figure 9.36

Left (Fig 9.36) comprises a staircase with open treads and cold-to-touch stainless steel handrails, which are not considered accessible. Fig 9.37 shows more accessible stairs with contrasting, textured nosings, closed treads and plasticised handrail.

Figure 9.37

The legal minimum requirement for accessible internal staircases or steps refers to the requirements for external access steps. However, it stipulates that it is not necessary to place corduroy textured surfaces at the top or bottom of the stairs internally. It essentially defaults to Building Regulations Approved Document Part K: *Protection from falling, collision and impact*[2] with the added recommendation for stair widths ('going') to be 300 mm but between 250 mm and 400 mm.

The overlap of stair nosings should be no more than 25 mm, and nosings should colour contrast with the risers and treads with a width of between 55 cm and 65 cm. Single steps should be avoided and replaced with a ramp; stair risers should also not be open with the minimum distance between enclosing walls being 1200 mm and 1000 mm between handrails. The number of risers to a flight is limited to 12 but up to 16 is permitted in exceptional circumstances. Landings with a permanently firm surface should be 1200 mm deep with a maximum gradient of 1:60.

The handrails to internal stairs should be placed 900–1000 mm above the pitch line of the stairs. These should be continuous, colour-contrasting, non-reflective and made from materials that do not become excessively cold to touch. They should extend 300 mm past the first and last tread and prevent clothes getting caught on them. A second low-level handrail should be installed at a height of 600 mm above the pitch line.

Best practice largely adopts the same or similar requirements for internal stairs as the legal minimum, but it does permit wider tread widths and also a maximum of 20 risers per flight. Level changes of up to 300 mm should accommodate a ramp, and over 300 mm a ramp or stairs can be used as per the limits of ramps. Nosings should be textured as well as colour contrasting, and the material choice to stairs should be non-reflective and slip resistant when wet; deep-pile carpet is also specifically discouraged. Best practice recommends lighting levels of 100 Lux at tread level and also suggests that spotlights, inappropriate wall or low level, are not used. Best-practice guidance concerning handrails is similar to the legal minimum and also recommends a second handrail to public buildings.

FACT FILE: Ramps / Slopes / Handrails

Figure 9.38

Fig 9.38: Illustrates an internal ramp to a historic, listed building that encompasses contrasting floor and walls as well as a handrail to both sides of the ramp. Noted is no contrast in the floor colour to the landings, and the handrails appear to be quite wide and not easily gripped. Accessible handrails should be colour contrasting, easy to grip, not cold to touch and designed not to catch clothes (Fig 9.39).

Figure 9.39

It is recognised within the legal minimum requirements that internal ramps are permitted to provide vertical circulation. The design criteria of these defaults to Building Regulations Approved Document Part K and stipulates the following:

* minimum width of 1500 mm between walls, upstands or kerbs;
* free from permanent obstruction with handrails to both sides;
* landings 1200 mm long at the top and bottom free from door swings;
* intermediate landings should be 1500 mm long and free from door swings;
* lengthy ramps or those of three flights or more require 1800 mm wide passing places;
* slip resistant surface and colour contrast with the landings;
* 100 mm visually contrasting upstand / kerb to the open side of the ramp;
* stepped alternative to the ramp where the vertical rise is greater than 300 mm;
* maximum gradient of 1:12 and clear height to soffit a minimum of 2.000 m;
* handrails to one side of the ramp where this is less than 1000 mm wide and to both sides where the ramp is wider;
* no handrail is required where the ramp height is less than 600 mm; and
* firm, easy gripped handrails between 900 mm and 1000 mm above the ramp.

Best practice aligns with the legal minimum requirement on several items but also recommends landings 1500 mm long every 500 mm of vertical distance travelled with the positioning of ramps in close proximity to stairs. Ramp lengths between landings are limited to 10.00 m in length or 500 mm in vertical distance; above 2.00 m of vertical height, a lift or platform lift should be installed instead of a ramp. Gradients should be between 1:12 and 1:20 with a cross fall of no more that 1:50, which also applies to landings to facilitate the draining of surface water. The frictional characteristics of the non-slip surface materials should be similar between the ramp and landings, deep-pile carpet, and shiny surfaces causing glare should be avoided. It is also not recommended to place tactile (corduroy) surface as the top and bottom of the ramp. Best practice collectively addresses handrails to ramps and steps simultaneously; these are recommended to be 900–1000 mm above the pitch of the ramp with a lower 600 mm handrail to publicly accessible buildings. Handrails should be easily gripped with no sharp edges, have sufficient frictional resistance and use colour contrasting. They should extend 300 mm passed the upper and lower ends of the ramp and be designed not to catch clothing. They should be sufficient in strength to resist impact from mobility scooters and be load bearing.

FACT FILE: Lifts and Lifting Devices (Landings and General Principles)

Figure 9.41 Visually contrasting lift signage and contrasting doors / walls.

Figure 9.40 Visually contrasting, tactile, Braille buttons.

According to the legal minimum requirements for lifts and lifting devices, these should have a 1500 mm x 1500 mm space in front of the doors and controls or a 900 mm wide space straight on an access route.

- lift buttons or controls should be placed between 900 mm and 1100 mm above finished floor level and a minimum of 500 mm from any return wall;
- tactile buttons should be colour contrasted against their mounting plate and the plate contrasted against the wall or surface;
- lift landings and car doors are contrasting against adjoining walls;
- the floor finish of the lift / lifting device should not be dark in colour and the surface should have similar or more frictional resistance than the landing;
- a handrail should be installed in the lifting device 900 mm above the floor and not blocking the mirror or controls; and
- an emergency call system should be installed.

Best practice addresses access to lifts by firstly recognising the need to signpost the location of these at the building entrance with a need to explicitly indicate accessible lifts as there may be instances where lifts are not accessible. These should be located at the same level as the entrance or accessible via a ramp. The landing space in front of the lift should be 1500 mm x 1500 mm and clear of any door swing; the provision of lighting should be 100 lux at floor level.

The research findings have identified that to those with a physical and mental impairment, the presence of lifts is an essential adjustment and not a reasonable adjustment. Those with disability prioritise the requirements for lifts in the following order:

- visually contrasting signage (1.31);
- tactile buttons (1.38);
- visual contrasting lift doors (1.46);
- sufficient lighting to lift lobbies (1.50); and
- non slip and frictional qualities of floor surfaces of the (1.75)

FACT FILE: Passenger Lifts

Figure 9.42   Figure 9.43

It is recognised that the Equality Act 2010 does not stipulate the placement of a lift to any commercial provider of goods, services or employment. However, new build or significant renovation projects may stipulate the placement of a lift. Accordingly, the legal minimum requirement for passenger lifts requires this to comply with specific lift regulations and British standards. Vertical circulation between levels should be accessible from all storeys, and the minimum dimensions of a lift car should be 1400 mm deep and 1100 mm wide. These minimum dimensions do not permit wheelchair users to turn around, and where this is not possible, a wall-mounted mirror should be provided to aid the reversing process.

Power-assisted doors should have 800 mm door clearance and be equipped with a timing device as well as sensors to prevent this closing on the user. The placement of the internal lift controls should be 900–1200 mm above finished floor level and a minimum of 400 mm from any return wall. There should be both audible and visual indication to signal destination arrival in both the lift car as well as the landing. Glazing to lifts should include the necessary manifestation, and where lifts are designated as emergency evacuation devices, these should conform specifically to British Standard BS5588-8.

Best practice aligns mostly with the legal minimum requirement but further recommends:

- lift door width of 800 mm for existing buildings but 900 mm or new build;
- the lift car floor should contrast with the floor of the adjacent landing;
- glass floors should be avoided and at least one solid or opaque wall should be present;
- lighting levels of 100 lux should be present to the control panels and 1.00 m above the floor without causing pools of light or shadow; and
- large lifts (2000 mm wide and 1400 mm deep) should have duplicate controls on opposite sides.

All of the minimal legal requirements and best practice are considered to be an essential adjustment according to the findings of the research and not deemed just 'reasonable'.

FACT FILE: Platform Lifts

Figure 9.44

Platform lifts are a good and low cost alternative to traditional lifts but are typically limited to providing vertical circulation between two levels, unlike lifts, which are required for multi-storey buildings. The legal minimum requirements for these transport devices stipulate conformity with relevant standards or notifying bodies and that they are limited to a travelling distance of 2 m when these are open devices and not enclosed. Platform lift dimensions vary dependant on whether they are non-enclosed and intended for unaccompanied use (800 mm wide x 1250 mm deep), 900 m x 1400 mm for enclosed unaccompanied use or 1100 mm x 1400 mm for enclosed accompanied use. Other requirements include:

- max speed of 0.15 m/s;
- permanent pressure control buttons and control panel at 800–1100 mm (min 400 mm from flank wall);
- landing control buttons placed between 900 mm and 1100 mm (min 500 mm from flank wall);
- door width of 800 mm except for 1100 mm wide lifts which should be 900 mm;
- doors contrast with adjoining walls;
- visible instructions;
- audible and visual arrival announcement indicator; and
- glazed areas include manifestation.

Best practice recommends that platform lifts are equipped with emergency communication devices. The floor of the lifting platform should contrast with the landing, and the transition between the landing and platform is level (ramped transition should be avoided). Enclosed platform dimensions should be 1100 mm x 1400 mm, except in exceptional circumstances where 900 mm x 1400 mm is permitted.

FACT FILE: Wheelchair Stair Lifts

This is not covered by the legal minimum standards but is detailed in best practice with the notion the wheelchair stair lifts are not considered an acceptable installation to new buildings. These are an example of a retrofitted, 'reasonable adjustment', which is seen as an acceptable compromise, but only if a lift or platform lift cannot be installed.

The biggest concern regarding the placement of a wheelchair stairlift to an existing staircase is the risk of these compromises to means of escape from the building. Accordingly, best practice recommends that such installations should be placed in the secondary staircase of a building where there are two staircases, allowing for the primary means of escape to remain clear.

There is an applicable British Standards (BS EN 81-40) relating to the specification of wheelchair stairlifts. This also addresses the installation of an alarm, as well as measures to ensure it cannot be used without authorisation.

An important consideration in the installation of a wheelchair stair lift is when this is in the 'parked' position and that it does not obstruct the means of escape or even present as an obstacle or hazard to those with sight loss or severe sight loss.

There is only one dimension of note in the best practice recommendations regarding wheelchair stairlifts and this is that the minimum distance between the folded down platform and opposite handrail should not be less than 600 mm

Signage and Communication

FACT FILE: Signs and Information

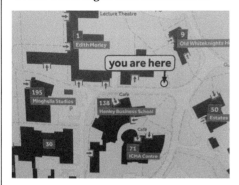

Figure 9.45   *Figure 9.46* Tactile, Braille signage.

There are no specific minimum legal requirements concerning signage, but this is addressed with best practice recommendations. This identifies that the effectiveness of signage and communication is determined by the location or placement of signs, as well as the accessibility, layout, and height. Orientation signage indicating *you are here* should be positioned, along the main access routes with the placement of signage not inhibiting or limiting the access route. Directional signage should illustrate logical routes, as well as escape routes and the positioning of both steps and ramps. There is a requirement to use universally recognised signs and symbols, using blue text for mandatory instruction, green for safety, yellow for hazard warning, and red for danger or emergency. With the exception of *push / pull* or warning signs, most door signs should be placed on walls and not doors; these should be placed to avoid reflection from both natural and artificial light.

The size of lettering, effective use of colour and contrast, as well as tactile lettering and symbols add significantly to inclusivity. Text entirely in capitals should not be used, and uniform, standardised fonts should be adopted. The size of lettering should be 150 mm minimum for long-range approach signage, 50–100 mm for directional signage, and 15–25 mm for room signs. Light-coloured text on a dark background is preferrable, although the key requirement for good contrast is a difference in light reflectance value between the lettering / symbols and background of 70 points.

Tactile pictograms and Braille signage should also be placed, and best-practice guidance indicates the height of embossed lettering with a preference for this over recessed or engraved lettering. Emphasis is placed upon the edge detail to the letters or pictograms, which should be slightly rounded but not half round. The positioning of Braille on directional signage should be identifiable by the blind or visually impaired through the use of a notch or tactile shape to the lefthand edge of the sign.

Best practice recognises the need for audible signage or information alongside tactile signage but does not detail specifics. It notes that invasive audible announcements can be stressful to hearing aid users.

FACT FILE: Assistive Listening Systems – Hearing Loops

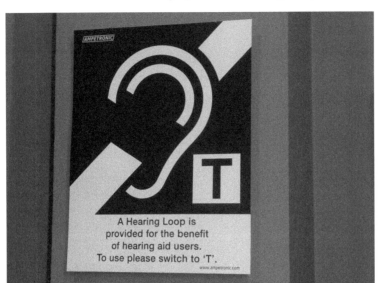

Figure 9.47

Best practice addresses a number of assistive listening technologies for deafness including:

- hearing loops;
- infrared systems;
- radio and Wi-Fi systems; and
- inductive couplers.

Hearing loops are one of the more common applications and rely upon a microphone at source to pick up the audio. This can be in a location such as a ticket desk at a railway station or at the podium of an auditorium. The microphone is connected to a wired induction loop in the immediate vicinity of the users, such as the other side of the ticketing desk or surrounding seating in the auditorium. The user then selects the *T* function on their hearing aid to tune into the signal and obtain effective sound, irrespective of background noise. This simple but effective technology is appropriate for a high number of users although other loops in close proximity can cause interference. Also, the presence of metallic building structure or components, as well as electrical equipment can distort the signal. These are a specialist installation and, accordingly, specialist advice should be sought.

Best practice recognises that infrared systems are typically used in cinemas or with theatre audiences, and they require use of headsets to be supplied by the service provider. The number of headsets made available should be 3%–4% of the maximum audience capacity, and these can have more than one channel to accommodate other features such as audio description. Infrared systems are a specialist installation, and it will be necessary to default to the advice or recommendations of the manufacturer or installer.

The use of radio systems requires the installation of a transmitter for the source sound or information and individual users with receivers to pick this up. This system is typically used for tour guides or in museums and lecture space. There is an obvious requirement for there to be sufficient numbers of receiving devices. This technology is considered to be quite old but also very effective. A more modern version of this is Wi-Fi communication, which utilises online technology and smart phones for user experience. This type of technology is becoming more widespread in assisting provision of inclusive environments.

FACT FILE: Fire Alert, Alarm and Means of Escape

Figure 9.49 Visual fire alarm.

Figure 9.48 'Refuge'.

The legal minimum requirement for fire alert, alarm and means of escape is contained in the broader advice given in Building Regulations, *Approved Document Part B Fire Safety*. This identifies the need for visual alarm for those with deafness. Best practice addresses this in more detail indicating that:

- the alarm should be audible and visual;
- audible sounders should not interfere with communication capability in refuges;
- areas of relative isolation such as toilets, bathrooms or individual offices should be fitted with flashing beacons and vibrating devices additional to the fire alarm; and
- a large number of low output devices as opposed to a low number of high output devices should be installed.

Best practice recognises that vibrating devices can include those worn by a person and also that certain flashing frequencies can cause epileptic seizures, migraines and nausea.

Other audible alarm systems should be installed, but it is necessary to ensure these are not confused with fire alarm call points. Typically, these are two red pull cords installed to toilets and other isolated locations equipped with 50 mm diameter red bangles. One is positioned between 800 mm and 1000 mm, and the other 100 mm above finished floor level. This is connected to an indicator that can easily be heard or seen as a request for assistance.

In the event that it is not possible for a person with a disability to evacuate a building via the designated escape routes, it is necessary to foresee a refuge where they can wait for assistance. This is a safe space that meets the legal minimum requirements:

- there should be one refuge per level with direct access to the emergency stairs;
- these can be placed within staircases with a min dimension of 900 mm x 1400 mm and that they do not block or impede the escape route;
- these need to be fitted with an emergency voice communication system; and
- markings in blue text need to indicate the presence of the refuge and to keep clear.

Clearly under fire conditions or with the order to evacuate a building, staying put in a refuge space can cause anxiety, and it is normal to foresee other specialist equipment such as evacuation chairs for facilitate assisted escape.

Facilities and Sanitary Rooms

FACT FILE: Seating & Waiting Areas

Fig 9.50: A waiting area missing different seating heights.

Figure 9.50

While the legal minimum requirements recognise the significance of a building entrance area and reception, there is nothing stipulating the requirements for seating or waiting areas. This however is covered with best practice recommendations, and it is necessary to consider the different types of waiting areas. There are wating areas in the lobby space of offices that are often open space with attractive, comfortable seating, sofas and tables. Compare to this the 'traditional' notion of waiting areas associated with medical centres or transport hubs where there are often rows of seats organised in a more formal, structured layout.

Regardless of the layout, seating to waiting areas must pay attention to the seat heights with a recommendation for a variety of this to include levels of 380 mm / 480 mm / 580 mm, but if only one chair is used, then it should be between 450 mm and 480 m above finished floor level. Best practice also recommends quite a lot of detail concerning the placement of arm rests, as well as the space required to transfer from a wheelchair onto bench seating. The actual provision of accessible seating to waiting areas is not as simple as it might appear.

Waiting room seating should be accessible, and best practice identifies the necessary distances between rows of seats as well as the requirement to accommodate two adjacent wheelchair users. The surface of waiting areas should be level, and best practice suggests the building destination use is an important consideration, particularly with transport hubs and the need to foresee the provision of additional space for luggage. Typically, also with transport hubs are large horizontal travelling distances, and best-practice recommends seating every 50 m.

There is a recognition that seating may be within circulation areas, but this should not impede access with sufficient clear space; furthermore, seating should visually contrast with its surroundings to make this identifiable. Audible and visual communication should be foreseen within waiting areas, and this should include signage, as well as assistive hearing devices as previously detailed, including the ability for those with applicable hearing aids to switch to the *T* setting.

FACT FILE: Storage Facilities

There are no legal minimum requirements for storage facilities, and contained within best-practice guidance is evidence of why this embodies a comprehensive approach to inclusivity. Tangible examples of storage within the built environment for those with a physical or mental impairment relate to storage cupboards, cabinets and shelves associated with office work. Bookcases, shelving and racking could be envisaged to cover a number of applications or service provision. In summary, the best-practice recommendations contain a series of clear images illustrating access provision for both the ambulant disabled and wheelchair users.

The requirement to consider the adaption of shelving and storage facilities within the workspace is no doubt a reasonable adjustment. Most modern office furniture has the flexibility in design to accommodate the placement of shelves and drawers curtailed to individual users.

Some of the key dimensions for wheelchair users include the following:

- low-level shelf heights of 650 mm above finished floor level;
- high-level shelf heights of 1000 mm above finished floor level; and
- bookshelves or drawer pulls not lower than 400 mm above finished floor level.

There is a recognition that those who can stand up may have some restrictive movement, such as difficulty in bending or stretching and best practice recommends that the upper level of shelving should be 1500 mm and 750 mm for lower.

Some common key requirements for inclusive design include the use of colour contrast for projecting items of shelving or drawers and contrast of shelving against the background materials. Shelves and projection should also include edges that are not sharp and don't present a danger. Easy-grip handles with low resistant door mechanisms to cupboards should be foreseen, and the materials used for shelving and cupboards should not be reflective to minimise surface glare.

FACT FILE: ATMs and Cash Operated Devices

With a shift towards a 'cashless society', it could be questioned why there is a need to consider the accessibility of ATMs ('cashpoints') and other cash-operated devices, such as those requiring coin or even card operation. There is no requirement under the legal minimum stipulated in Building Regulations Part M concerning this. However, the access to ATMs, cash or card-operated devices and the absence of access provision may be something challenged under The Equality Act 2010.

One of the obvious difficulties for gaining access to cash or card-operated devices concerns wheelchair users and how these devices are accessible from a seating position. Some of the key requirements identified in best practice concerning these devices are:

- unobstructed access routes;
- clearly signposted;
- well lit and not overlooked (important for ATMs);
- display screens should be shaded from natural light to avoid glare or reflection;
- clear and illuminated signage should be evident indicating the presence of ATMs; and
- external ATMs should ideally be protected from the weather.

For wheelchair users accessing ATMs and other cash or card-operated devices, it is recommended that there is a 500 mm knee recess under at a height of 70 mm to enable these to be operated from the front. Although not a physical building requirement, the service provider of ATMs should ensure that the text utilised by the devices is clear with sufficient typeface size to facilitate ease of use. The use of universal signage and pictograms is recommended.

FACT FILE: Windows & Their Operation

There is a recognition that wheelchair users and those with disability who spend significant amounts of time sitting down should be able to look out of windows and operate these as is expected for those without disability. While best practice makes a number of recommendations on implementing design considerations, it should also be noted that the windows to commercial buildings are also covered by other specific standards, as well as building regulations. With current new build office accommodation and the need to maximise energy efficiency, there is a drive to ensure that buildings are sealed without the possibility to open windows, thus allowing the complete internal climate to be controlled.

Where there are other buildings where window design incorporates the ability for these to be opened, it is recognised that windows may be either casement type, opening either inward or outward or pivoted, meaning they are hinged centrally in the horizontal or vertical plane. Windows can also be either sliding sash opening vertically or horizontally and, in some cases, casement windows have a 'tilt and turn' mechanism allowing them to be partially opened.

Concerning windows, best practice guidance recommends:

- clear view with window transoms (horizontal framed sections) positioned below 800 mm and above 1200 mm;
- opening mechanisms at 800–1000 mm but with locking facility to prevent these opening more than 100 mm to prevent fall risk from upper floors;
- opening windows at ground floor level next to an access route should be restricted to 100 mm opening or external barriers should be placed to prevent impact risk with pedestrians;
- lever-opening mechanisms are preferred over knobs, and these should be operable with a closed fist; and
- specific torque levels or resistance required to operate opening mechanisms.

Further guidance is provided for the placement, function and operation of automatic window-opening devices. This includes the type of device as well as the location of wall-mounted control panels, remote-controlled opening mechanisms, and if these are single-press controls or require continuous pressure. One common theme with the placement of window controls is that they should contrast against their background.

FACT FILE: Outlets / Sockets / Switches

The legal minimum requirements for the placement of sockets and outlets concerning the accessibility of buildings identifies a number of key design considerations. Full inclusivity needs to consider all disabilities; therefore, one of the key initial requirements is for switches, sockets and outlets to contrast visually against their background. There is a need for consistency in the positioning of these:

- sockets, outlets, telephone point and TV socket heights should be 400–1000 mm;
- permanently wired appliances 400–1200 mm unless appliances are higher;
- switches and control that require specific hand movements 750–1200 mm;
- push button controls for limited dexterity at a height of not more than 1200 mm;
- emergency pulls cords are red with 2 × 50 mm diameter bangles 100 mm and 800 mm – 1000 mm above finished floor level;
- controls that need close vision are at heights of between 1200 mm and 1400 mm;
- sockets are placed a minimum of 350 mm from room corners;
- light switches for general public use are push pads aligning to door handles between 900 mm and 1100 mm;
- pull light cords are 900–1100 mm, fitted with 50 mm bangle visually contrasting but not red;
- switches, outlets and controls should not require the use of both hands simultaneously;
- switched sockets indicate when they are in the 'ON' position;
- mains isolator switches or circuit breakers indicate ON / OFF positioning; and
- front plates contrast visually against their backgrounds.

Best practice aligns largely with the legal minimum standards with some additional recommendations such as the need for clarity with main isolation switches indicating the on and off settings. While this is normally indicated using green (on) red (off), such colours are not acceptable as they can be confusing for those with visual impairment or colour blindness. Furthermore, 'Break glass' fire alarm call points should be positioned between 1000 mm and 1200 mm above finished floor level. The research has identified that the height of sockets, switches and outlets is the most 'essential' adjustment in this category for accessing commercial property, goods, services and employment. Compliance with best practice is considered to be more than a reasonable adjustment.

FACT FILE: Audience and Spectator Facilities

The legal minimum requirement for audience-spectator facilities stipulates that the route to wheelchair spaces is accessible by wheelchair users; where there is stepped access, this needs to comply with the previously specified requirements for handrails.

Figure 9.51

Concerning the minimum number of permanent and removable spaces for wheelchair users this is:

- 1% of seating capacity permanently allocated as wheelchair space for seating capacities up to 600 people or the remainder of removeable seating to deliver a total of six spaces;
- for capacities of between 600 and 10,000, there should be 1% provision for permanent wheelchair access with additional removeable seating optional; and
- see guidance for 'Accessible Stadia'[3] where seating capacity is > 10,000.

Some of the wheelchair seating should be arranged in pairs and always with at least one standard seat adjacent to it. Where there is more than one wheelchair seating, these should be placed at different locations to give a range of different views. Access to spaces should be 900 mm wide, and the spaces themselves should measure 900 mm in width with a depth of 1400 mm; the floor to wheelchair spaces should be horizontal. The legal minimum requirements note that some seats are able to provide space for an assistance dog in front or under the seat, and arm rests to the ends of rows should also be removeable. There are specific requirements for wheelchair spaces on stepped terrace floors.

   Best practice for audience seating aligns with most of the requirements stipulated as legal minimum but recommends greater amounts of wheelchair spaces (3%) including removeable seats plus six accessible seats or 4% of the capacity (whichever is greatest) where the total seating capacity is 600. For venues with seating capacity greater than 10,000, a total of 3% should be dedicated to wheelchair users and 4% as accessible seating. Best practice also goes further to detail the need for visual contrast to access routes, as well as seating and the need for sight lines to facilitate lip reading as well as lighting. Access routes to raked floors (stepped terracing) should encompass handrails and wheelchair spaces should also be level with protective barriers or railings. Concerning audience and spectator facilities, best practice goes significantly above and beyond the legal minimum requirements. For full inclusivity at these venues, it is necessary for service providers to consider adopting best practice.

FACT FILE: Lecture Space and Conference Venues

Figure 9.52

Figure 9.54

Figure 9.53

Renovation and retrofit out of a 1970s lecture space to include the provision of a platform lift (Fig 9.52) and a signposted power-assisted entry door (Fig 9.54), which leads directly to two end-of-row accessible spaces equipped with power supplies for assistive technology devices (Fig 9.53). The space has been fit out with acoustic internal wall linings and a hearing loop.

The legal minimum requirement recognises where facilities are used for lecture space or conferences wheelchair users should be able to access the stage or podium via a ramp or lifting appliances and overall, the space should be equipped with a hearing enhancement system.

Best practice concerning lecturing and conference facilities goes into significantly greater depth which includes specific dimensional requirements for ancillary equipment such as lecterns and overhead projectors or keyboard / computer facilities. There is a recommendation for lecterns to be adjustable in height with appropriate lighting to also encompass the illumination of the speakers face to facilitate lip reading. Concerning space dedicated to study, the heights of fixed desks should be between 730 mm and 750 mm with a clear height under the desks of 700 mm with a view, where feasible to have adjustable desks. There should be sufficient space between desks to provide and access route and noted is the recommendation to foresee power sockets to recharge mobility devices and assistive technology equipment such as laptops or tablets.

Signage, lighting, and lettering of projected images should meet with that prescribed within the guidance for signage and information. Best practice recommends the use of visual contrast for the placement of seating to conference facilities, and with respect to acoustics, the materials used as finishes to the floors, walls and ceilings should have appropriate sound absorption qualities.

FACT FILE: Sanitary (Toilet) Accommodation – General

Figure 9.55   *Figure 9.56*

(left) Epitomises everything that is wrong with accessibility: the only accessible toilet is located on the middle floor of a five-storey, 10,000 m^2 building. Fig 9.56 (right) is good inclusive design with the accessible toilet in the immediate vicinity of non-accessible toilets.

Probably one of the most visible attempts to engender an inclusive environment is the placement of accessible toilets. These are typically present to most public spaces, public buildings, tourist attractions, places of work and some service providers. Several general principles apply regarding the provision of accessible toilets, and these are not obliged to be present in every building offering access to goods or services. The legal minimum requirement stipulates that at least one accessible, unisex toilet is provided where there are sanitary facilities for use by customers, visitors to the building or those working on the premises. Accordingly, there is an argument that certain premises such as those not serving food or drink should not foresee accessible toilets as they would not provide customer toilets. The counter argument is that even the employment of staff within a commercial building who have access to a toilet should also be able to use accessible facilities. Best practice does not entertain the notion that there is a difference in service providers but appears to adopt the position that there should be accessible unisex facilities to all buildings. By indicating that those with disability should not be disadvantaged in anyway concerning the provision of accessible toilets, best practice simply appears to recognise that wherever there are toilet installations, these should be fully accessible.

   The legal minimum requirements concerning sanitary accommodation recognises not just the provision of at least one unisex, accessible toilet but also the need to provide one WC cubicle in separate-sex toilet accommodation for the use of ambulant disabled people. The legal minimum goes on to further oblige one enlarged cubicle in separate-sex toilets where there are four or more cubicles additional to a cubicle for ambulant disabled. The legal minimum requirements also address the need and criteria for the placement of a *Changing Places* toilet, which will be addressed in a later fact file.

It is necessary to put some context into both the legal minimum requirement as well as the guidance recommended by best practice and the need to mostly adopt retrospective measures to providing accessible toilets within the existing built environment. While there appears no excuse why accessible sanitary provision cannot be installed to new build projects or those undertaking significant renovation, challenge and compromise is always a consideration with retrofit out. The research findings identified that all of the items considered to equate to the legal minimum requirements for the provision of sanitary accommodation are 'essential' adjustments and not just 'reasonable adjustments'. The highest priority evident in the research data is for a unisex accessible toilet cubicle to be placed at every location where sanitary facilities are present (1.32).

FACT FILE: Unisex Sanitary (Toilet) Accommodation – Specific

Figure 9.58 Lever operated tap.

Figure 9.57 Colour contrasting fixtures.

The general principles stipulate within the legal minimum requirements for toilet provision and the need to place these adjacent to standard sanitary facilities as well as being close to the building entrance. Further detailed in the text and also with annotated plans are specific unisex accessible requirements that include:

- alternate left-hand and right-hand transfer to be foreseen with toilets on different floors;
- where only one toilet is installed, this needs a width of 2.00 m to accommodate a standing-height wash hand basin additional to the finger rinse basin;
- toilets to be located on access routes that are direct and free from obstruction;
- doors preferably open outwards and are fitted with a horizontal closing bar;
- travelling distance to toilets should be no more than 40 m on the same floor or more than 40 m over one vertical travelling distance where a lift device is in place;
- specific design requirements (heights and positioning of equipment) as detailed;
- emergency alarm (visual & audible) operated internally by a pull cord as detailed;
- specific heights of the WC pans and conformity with BS EN 997:2012; and
- the flush mechanism for cisterns is positioned to the open or transfer side.

Best practice makes additional recommendations detailing specific advice on the types of wash hand basins, lever-operated taps with clear markings and heat control of water to 43°C. There are further recommendations to box in hot pipework provided this does not alter clear widths. Key dimensions are also recommended with the diameter and positioning of grab or support rails as well as the provision of coat hooks, towel rails and shelves. Best practice details the rationale for the outward opening of entry doors to ease emergency access as well as the recommendation when only inward opening doors are foreseen. Importantly where outward opening doors are positioned into corridors, these should be recessed as deep as the width of the door leaf.

Additional recommendations are made on the colour contrast of fixtures and fittings as well as avoiding shiny floor surfaces in favour of non-slip materials. The provision of an emergency alarm aligns with the legal minimum requirements plus the need for a reset facility adjacent to the toilet seat if this is activated by mistake. Best practice recognises the need to provide lighting levels of 100 lux at floor level and that this can be provided by either automatic, sensor controlled lighting or pull switches.

FACT FILE: Ambulant Disabled Sanitary (Toilet) Accommodation

Figure 9.59 An attempt at installing an ambulant disabled WC, which does not appear to be fully compliant.

The legal minimum requirements state that ambulant people with disability should have the opportunity to use a toilet within separate-sex sanitary rooms. These are effectively enlarged cubicles with a width of 800 mm, and these are equipped with horizontal grab rails either side of the WC pan, an additional vertical grab rail and clothes hook positioned at 1400 mm above ground floor level. The door should be outward opening, but where this is inward opening, there must be a 450 mm manoeuvring space between the door and pan. The cubicle can be further enlarged to a width of 1200 mm for those needing extra space with additional grab rails to the side and rear wall as well as a fold-down baby-changing table or shelf. Noted is a subtle difference in the wording for an ambulant disabled cubicle and an ambulant compartment and wheelchair accessible washroom. In essence, the 'washroom' should be the same fit out as a unisex toilet, and there needs to be a wash hand basin with its rim set to a height of between 720 mm and 740 mm. Where a urinal is fitted, this should include a rim height of 380 mm with 2×600 mm vertical grab rails fixed either side of this at a height of 1100 mm.

Best practice recommendations for ambulant disabled cubicles align with the legal minimum but additionally suggest the cubicle can be between 800 mm and 1000 mm wide; furthermore, an additional coat hook placed at 1050 mm above finished floor level should be foreseen. There is a recommendation in adopting best practice to avoid inward-opening doors, and there is an acceptance that this may only be acceptable to existing buildings where there is no alternative.

Best practice suggests that toilet flushes should be operated manually by a 'spatula' type lever located between 800 mm and 1000 mm above finished floor level positioned on the open / transfer side of the toilet. Toilet seats should be at a height of 480 mm for wheelchair users and have sufficient fixing strength to permit the relevant lateral force associated with transfer from a wheelchair. Accordingly retaining buffers or seat lugs can offer greater stability and toilet seat covers or lids should as well as open or 'gapped' toilet seats not be used.

FACT FILE: Shower Rooms

The legal minimum requirement for the provision of shower rooms stipulates that where showering facilities are provided in a commercial building, at least one of these should be accessible. It should be noted that shower facilities are also considered to be a standard installation to commercial office buildings, and accordingly, the principles on the provision of accessible toilet facilities should be applied to the provision of shower installations. The criteria stipulated as the legal minimum include:

- a choice of left-hand and right-hand transfer;
- tip up (not spring action) non-slip seats and drop-down support rails;
- emergency assistance pull cord and alarm as foreseen with toilets;
- shower curtain operable when seated;
- slip resistant floor that is free draining;
- shelf reachable from the seated position for toiletries;
- shower controls between 750 mm and 1000 mm above finished floor level;
- and facilities for limb storage for amputees.

Best practice aligns in part with the legal minimum requirements but recommends the number of accessible shower facilities should reflect the number of disabled people likely to use the facilities. In reality and outside the provision of accessible rooms in hotels or other accommodation, the numbers of shower cubicles for an office is a relatively small number in comparison with toilet facilities. It is believed that there is a small but committed number of building users likely to shower at work. To facilitate the needs to disabled users, occupants or visitors, the best way to accommodate accessible shower facilities, where space may be limited, is to provide these as unisex, accessible cubicles throughout. Best practice also refers to a series of dimensions illustrated within a diagram but specific additional requirements to the legal minimum include:

- two tip-up seats, one wet and one dry with non-slip finish should be provided;
- lever-operated thermostatic mixer valve with clear markings and maximum temperature limited to 43°C;
- and a lighting level 300 lux at floor level to the shower cubicle.

FACT FILE: Accessible Baby Changing Facilities

The legal minimum requirements concerning sanitary accommodation does not address accessible baby-changing facilities, although it is touched upon in the provision of enlarged unisex and same-sex accessible toilet cubicles.

Best practice recommends that where baby-changing facilities are installed, these should be accessible; accordingly, it is not appropriate for such facilities to be located within an existing unisex accessible toilet cubicle as this should always be available for users needing to access a toilet independent of baby changing. However, there is a recommendation to install a toilet within the accessible baby changing facilities and explicit reference is made to family toilet installations, as per British Standard BS 6465-2.

The base requirements for an accessible baby-changing facility, as detailed in best practice, is that the room measures 2.00 m x 2.00 m, specific requirements detailed in the provision of best practice include:

- a wall mounted fixed or adjustable baby-changing table with a clear height to the underside of 700 mm and maximum height of 750 mm;
- wash hand basin with a rim height or between 720 mm and 740 mm;
- soap dispenser and automatic hand dryer with their undersides between 800 mm and 1000 mm;
- full length mirror with its bottom edge 600 mm above finished floor level;
- a nappy vending machine with controls no higher than 1000 mm above finished floor level;
- a sanitary disposal bin, preferably recessed into the wall; and
- a chair if a fixed height changing table is installed.

FACT FILE: Changing Places Toilets

Figure 9.60   Figure 9.61

The legal minimum requirements for *Changing Places* toilets details and defines this important accessible sanitary facility. There is a recognition that changing places toilets are designed to facilitate the sanitary requirements of a disabled person with complex or multiple impairments which may require up to two assistants in the process. There is explicit reference to the individual organisation **Changing Places** who are experts in the installation of these facilities.[4] The legal minimum requirement also defaults specifically the best practice and the relevant information contained in BS 8300-2:2018.

The legal minimum requirement stipulates at least one changing places toilet should be provided where there is a building providing assembly, recreation, and entertainment with a capacity of 350 persons or more. The same requirement applies where there is a collection of small buildings onsite, such as zoos, theme parks and sports venues with a capacity of 2000 or more persons. At least one changing places toilet is obliged for shopping centres, shopping malls or retail parks where the gross floor area is 30,000 m^2 or more and also for retail premises having a gross floor area of 2500 m^2 or more. Additionally, the presence of changing places toilets is stipulated for sports and leisure buildings with a gross floor area of 5000 m^2 or more, as well as hospitals and primary care centres and crematoria or cemetery buildings. There is further specific definition of assembly, recreation and entertainment buildings. The provision of changing places facilities for education buildings is supposed to be separately addressed by the Department for Education.

As expected, best-practice recommendations for changing places toilets details a significant quantity of prescriptive requirements with accompanying diagram. In summary, best practice stipulates that the toilet cubicle should be 3.00 m wide and 4.00 m long with a ceiling height of 2.40 m. A clear door width of 1000 mm and level threshold should be foreseen with a single leaf outward opening door and a turning space of 1800 mm to the internal side of the entrance door. An overhead tracked hoist system with a safe working load of 200 kg should be placed and a wall-mounted or mobile changing bench also with a working load of 200 kg should be foreseen, which should be easily cleanable and allow for a shower facility to be installed.

FACT FILE: Individual Rooms – Bedrooms

The legal minimum requirement for sleeping accommodation obliges there to be one wheelchair accessible room for every 20 provided, with clear door widths as identified in Part B of the Approved Documents (Building Regulations). Swing doors to built-in wardrobes or cupboards should have 180° function, and sliding doors should have easy-grip, contrasting handles. Opening windows should have handles situated 800–1000 mm above finished floor level and easily operable (see earlier fact file: 'Windows and their operation').

Accessible rooms should be situated on accessible routes with access to all other facilities in the building. The standard of amenities in accessible rooms is the same as other bedrooms and designed to allow users to have a choice of location. Door entry and bathroom or shower facilities should be provided as already detailed in the legal minimum requirements. Wide angle viewers to entry doors should facilitate use by people standing or sitting at a height of between 1050 mm and 1500 mm. The size of a wheelchair-accessible bedroom should have sufficient (1500 mm x 1500 mm) turning space either side of the bed to facilitate independent transfer. Where there is the provision of a balcony, it should have compliant doors with a level threshold and no horizontal transoms between 900 mm and 1200 mm above finished floor level. Concerning balconies, there should be no permanent obstruction 1500 mm back from the doors.

There is an obligation to provide a visual fire alarm in all accessible bedrooms, as well as emergency-assistance alarms activated by a pull cord with the alarm signal outside of the room, which can be easily seen and heard to enable assistance or a central control point.

Best practice aligns largely with the legal minimum requirements but recommends on top of 5% of accessible rooms to hotels, 1% of all rooms should have a tracked hoist system. A clear space of 1500 mm x 1500 mm for turning and transfer from the side of the bed is required. This can be increased to 2100 mm x 2250 mm for the use of a mobile hoist. Where there is more than one accessible bedroom, this should incorporate a choice of shower or bath as well a choice of left-hand or right-hand transfer to the sanitary facilities. Best practice also addresses the provision of student accommodation recommending one accessible room or 4% of total room allocation as well as 1% for the provision of a tracked hoist system with the same specific features as detailed for hotel rooms. Accessible bedrooms and student rooms should also have obstruction free accessible routes, appropriate means of escape and be located close to lifts.

Notes

1   Tagg (2024)
2   Approved_Document_K.pdf (publishing.service.gov.uk)
3   Accessible stadia – Sports Grounds Safety AuthoritySports Grounds Safety Authority (sgsa. org.uk)
4   Changing Places Toilets (changing-places.org)

References

Bsigroup.com. (2018). *BS 8300-1:2018. Design of an accessible and inclusive built environment. Buildings. Code of practice.* [online] Available at: https://knowledge.bsigroup.com/products/design-of-an-accessible-and-inclusive-built-environment-external-environment-code-of-practice?version=standard [Accessed 14 January 2024].

Bsigroup.com. (2018). *BS 8300-2:2018. Design of an accessible and inclusive built environment. Buildings. Code of practice.* [online] Available at: https://knowledge.bsigroup.com/products/design-of-an-accessible-and-inclusive-built-environment-buildings-code-of-practice?version=standard [Accessed 14 January 2024].

Changing Places. (n.d). *What are changing places toilets?.* [online] Available at: https://www.changing-places.org/ [Accessed 14 January 2024].

HM Government. (2013). *The Building Regulations 2010, Approved Document Part K, Protection from falling, collision and impact.* [online] Available at: https://assets.publishing.service.gov.uk/media/60d5bdcde90e07716f516cfd/Approved_Document_K.pdf [Accessed 14 January 2024].

HM Government. (2020). *The Building Regulations 2010, Approved Document Part M, Access to and use of buildings, Volume 2 – Buildings other than dwellings.* [online] Available at: https://assets.publishing.service.gov.uk/government/uploads/system/uploads/attachment_data/file/990362/Approved_Document_M_vol_2.pdf [Accessed 14 January 2024].

Sports Grounds Safety Authority. (n.d). *Accessible stadia.* [online] Available at: https://sgsa.org.uk/accessible-stadia/ [Accessed 14 January 2024].

Tagg. (2024). Inclusive Environments: Access to Commercial Property, Goods and Services - Disability Specific Access Requirements. *Journal of Building Survey Valuation and Appraisal – Volume 13.* Henry Stewart Publications.

Index

Note: Page numbers in *italics* refer to figures.